WORLD IN PROCESS

SUNY series in
Constructive Postmodern Thought
David Ray Griffin, editor

David Ray Griffin, editor, *The Reenchantment of Science: Postmodern Proposals*
David Ray Griffin, editor, *Spirituality and Society: Postmodern Visions*
David Ray Griffin, *God and Religion in the Postmodern World: Essays in Postmodern Theology*
David Ray Griffin, William A. Beardslee, and Joe Holland, *Varieties of Postmodern Theology*
David Ray Griffin and Huston Smith, *Primordial Truth and Postmodern Theology*
David Ray Griffin, editor, *Sacred Interconnections: Postmodern Spirituality, Political Economy, and Art*
Robert Inchausti, *The Ignorant Perfection of Ordinary People*
David W. Orr, *Ecological Literacy: Education and the Transition to a Postmodern World*
David Ray Griffin, John B. Cobb Jr., Marcus P. Ford, Pete A. Y. Gunter, and Peter Ochs, *Founders of Constructive Postmodern Philosophy: Peirce, James, Bergson, Whitehead, and Hartshorne*
David Ray Griffin and Richard A. Falk, editors, *Postmodern Politics for a Planet in Crisis: Policy, Process, and Presidential Vision*
Steve Odin, *The Social Self in Zen and American Pragmatism*
Frederick Ferré, *Being and Value: Toward a Constructive Postmodern Metaphysics*
Sandra B. Lubarsky and David Ray Griffin, editors, *Jewish Theology and Process Thought*
J. Baird Callicott and Fernando J. R. de Rocha, editors, *Earth Summit Ethics: Toward a Reconstructive Postmodern Philosophy of Environmental Education*
David Ray Griffin, *Parapsychology, Philosophy, and Spirituality: A Postmodern Exploration*
Jay Earley, *Transforming Human Culture: Social Evolution and the Planetary Crisis*
Daniel A. Dombrowski, *Kazantzakis and God*
E. M. Adams, *A Society Fit for Human Beings*
Frederick Ferré, *Knowing and Value: Toward a Constructive Postmodern Epistemology*
Jerry H. Gill, *The Tacit Mode: Michael Polanyi's Postmodern Philosophy*
Nicholas F. Gier, *Spiritual Titanism: Indian, Chinese, and Western Perspectives*
David Ray Griffin, *Religion and Scientific Naturalism: Overcoming the Conflicts*

12/6/00

To Derrick,

John Jungerman

WORLD IN PROCESS

Creativity and Interconnection in the New Physics

John A. Jungerman

FORE WORD BY

John B. Cobb Jr.

STATE UNIVERSITY OF NEW YORK PRESS

Published by
State University of New York Press, Albany

Printed in the United States of America

For information, address State University of New York Press, 90 State Street, Suite 700, Albany, N.Y. 12207

Production by Diane Ganeles
Marketing by Dana Yanulavich

Library of Congress Cataloging-in-Publication Data

Jungerman, John A.
World in process: creativity and interconnection in the new physics/John A. Jungerman; foreword by John B. Cobb.
p. cm.—(SUNY series in constructive postmodern thought)
Includes bibliographical references and index.
ISBN 0-7914-4749-9 (alk. paper)—ISBN 0-7914-4750-2 (pbk. : alk. paper)
1. Physics. 2. Cosmology. 3. Religion and science. I. Title. II. Series.

QC21 .J86 2000
530—dc21 99-086457

10 9 8 7 6 5 4 3 2 1

To my grandchildren
who embody my wonder
and who mark my place in the ongoing process of the universe.

Contents

Foreword

Professor John A. Jungerman has written two books in one. One contribution is a remarkably readable, up-to-date, account of the present state of knowledge in physics. It is also remarkably comprehensive, covering special and general relativity theory, light, quantum theory, high-energy physics, complex systems, and cosmology. Other scientists have undertaken to share with the general public the conclusions of physicists. But I know of no other account that is both so inclusive and so accessible. Those who want to become knowledgeable about the world as now presented in physics can find no better source.

The second contribution is that of bringing physics, philosophy, and theology together. Jungerman does this through process thought, especially through the philosophy of Alfred North Whitehead. He summarizes this in the first chapter and develops its theological dimension in chapter 8. In addition, in each of the chapters on a particular branch of physics, he demonstrates how twentieth-century developments have led in the direction of replacing notions of inert material substances with interrelated processes.

We process philosophers and theologians have long argued that scientific developments in the twentieth century call for this shift in our fields of study. Often, however, our appeal to physics has seemed to have little connection with the way most physicists actually think or even with the basic concepts and categories with which they work. Most physicists, in other words, have shown little interest in our efforts. As long as they can develop new theories and test them, few physicists feel the need to reflect about the philosophical assumptions or implications of what they are doing.

This situation may be changing. Much to the satisfaction of process thinkers, there has been a recent upsurge of attention given by unusually reflective physicists to process thought. This book is an outstanding expression of that new interest. It can also function to expand and deepen its intensity.

Like most physicists, Jungerman had not been especially concerned about philosophy during his many years of teaching. Nevertheless, when he encountered process thought, he quickly recognized its congeniality with what he had been teaching. The enthusiasm engendered by this recognition resulted in this book.

Jungerman's response to process thought reflects a healthy concern that one should have a coherent worldview that does not exclude religious values and concerns. Hence he deals not only with events and relations as providing a way of understanding both physics itself and the world it studies, but also with the cosmos as a whole and with God as understood in process thought. Hence the book can function as an introduction to process thought just as it serves to introduce lay readers to physics.

In the twentieth century, according to Jungerman's account, physicists have adapted the conceptuality inherited from the nineteenth century to deal with phenomena to which this conceptuality does not readily apply. For example, the concept of waves continues to play a very important role even though its root meaning requires a substratum, and there is no such substratum. The concept of particles continues to be widely used even though the original understanding of particles was bound up with inert matter. The combination of two concepts derived from substance thinking in the single notion of wavicle does not really go far enough to overcome this problem.

Perhaps in the twenty-first century, physicists will develop a conceptuality that fits better with the discoveries of the twentieth. Jungerman emphasizes the fact that physics itself is in process. He suggests that thinking in terms of a field of events can encompass what has been learned about particles and waves.

Perhaps when physicists have recast their theories in more appropriate concepts, this will lead not only to greater coherence within physics, but also to richer interconnections among the sciences. Certainly it can support the effort, so well begun by Jungerman, to bring science, philosophy, and theology into fruitful relations. Process thought, which is also in process, can be enriched and transformed by new scientific knowledge. Perhaps we can move past the current epoch of the fragmentation of thought into one in which the pursuit of information and wisdom can be reunited.

JOHN B. COBB JR.

Preface

I had just retired from teaching physics at the University of California, Davis, for forty years. It had been a gratifying career. For me physics has a basic simplicity and aesthetic appeal, and modern physics and cosmology are a new world of mystery, wonder, and even weirdness. An opportunity soon presented itself that enabled me to share these ideas by teaching a class on the new physics for seminary students. They came from diverse disciplines—sociology, philosophy, systems theory, and twelfth-century mysticism. The course went well enough—the students seemed fascinated by their new understanding. Yet for me, and probably for them too, the spiritual framework was vague and unsatisfying. Although I tried to share my feelings for the beauty and order of the physical world, I had no ready answer to the inevitable religious questions. Why is it so lawful? What does it all mean?

Toward the end of the semester I invited the president of the seminary, an expert in process philosophy and theology, to address the class. The vista of process thought that she presented was not only compatible with what I knew about modern physics and cosmology, but it was also much more comprehensive and included the divine in a natural, even necessary way. For me, this was a turning point. So began my search to bring together these two realms: the new physics and process thought. Combined they produce a new myth appropriate to our age.

A myth is the story we tell ourselves to answer the great questions: How did the universe arise and what is our part in it? Is there a place for divinity? What is the meaning of our lives? With the new discoveries in physics and cosmology in the last few decades, we have an entirely new picture of the microworld of nuclei and atoms as well as the macroworld of the cosmos. Process philosophy and theology are able to integrate this new knowledge with our religious quest.

Our times cry out for a rapprochement between science and religion. In the West during the Enlightenment each went its separate way, to the detriment of both and to the peril of our civilization. Religion became to some extent irrelevant to our technological world, and technology and science became amoral regarding their impact on society. The results of this split are painfully apparent in the world around us. A spiritual malaise pervades the industrialized nations. The environment is under great stress to satisfy our demands for more material needs. The disillusioned who seek escape in drugs plague our societies. A sense of direction and hope has been lost. We are desperately in need of relevant religious guidance, but it is often absent, for example, organized religion does not usually address ecocide, but it is an increasing problem wrought by population pressures and technological "advances." Science is often employed for military applications or in the service of large corporations motivated largely by a search for profit.

Alfred North Whitehead introduced process philosophy and theology in the 1920s. It offers a worldview that is profoundly spiritual but is also in accord with modern science at a fundamental level. Process thought is comprehensive—it is a metaphysics that attempts to illuminate all our experience. A world of events and becoming rather than of substance and permanence are at the basis of process thought. Fundamental to process thinking is that the universe is becoming ever more complex and novel. This is in accord not only with the biologic evolution of our planet but also with the majestic evolution of the cosmos itself.

I wrote this book to emphasize the compatibility of process thought with what we know about the physical world and about the evolution of the universe, and to look for connections between them. To accomplish this goal I introduce some of the fundamental ideas of modern physics and cosmology with explicit connections to the process view. For me this kind of "web" thinking is very different from the usual scientific mind-set where one looks for more detail in a narrow area.

The first chapter introduces process philosophy. I shall refer to its concepts a great deal in subsequent chapters concerning physics and cosmology as I strive to make the connections that provide a worldview that integrates science and process thought.

I discuss ideas from the special and general theories of relativity in chapter 2. These ideas are fundamental to our understanding of modern physics and cosmology—they are our tools for this quest. The ideas also illustrate how science itself is a process. They invite new ways of looking at the world.

Subsequent chapters consider the physical world, first of the very small—waves and particles (chap. 3), quantum mechanics (chap. 4), and particle physics (chap. 5), then that of human scale: self-organizing systems and chaos (chap. 6), and finally the vast universe (chap. 7). Chapter 8 gathers together information from the physical world that supports the case for divinity, gives the principal ideas of process theology, and shows their compatibility with that divinity. Chapter 9 presents a summary of the principal ideas from process thought and from the new physics and their relevance to the world of our daily lives. Since there is new vocabulary used both in the new physics as well as process thought, a glossary is provided.

On all levels in the physical world we shall find three fundamental characteristics: creativity, an openness with an active selection among alternatives, and interconnection. Science shows us these qualities in the physical world. These same qualities are emphasized in process metaphysics, which attempts to describe all of our experiences, including religion—hence we are led to a comprehensive view of our place and role in the universe.

Acknowledgments

In writing this book for the intelligent person who has little prior knowledge of either physics and cosmology or process thought, I am greatly indebted to several friends who have patiently read ongoing versions of the manuscript: Jay Atkinson, Janelle Curlin-Taylor, Lisa Dettloff, Priscilla High, Stanley Keleman, Calvin Schwabe, and Sandy Westfall. I would like to thank Raymond Coppock for his special editorial skills. Lorraine Anderson deserves special mention for her careful editing; there was particular emphasis on deleting technical material so that basic ideas are presented as clearly as possible.

I am especially grateful to Rebecca Parker, Mildred Payne, and Jay Atkinson for introducing me to process thought and to John B. Cobb Jr. for his suggestions after reading an earlier version of the manuscript and for his thoughtful foreword. William Beardslee and Robert Mesle also made special contributions to the chapters on process thought. I wish to thank Palmyre Oomen for her patient explanations of her formulation of the prehension of the divine consequent nature. I am indebted to Timothy Eastman for his detailed comments on the new physics presented and its relation to process philosophy and to Geoffrey Chew for his stimulating insights.

David Ray Griffin, editor of the SUNY series in Constructive Postmodern Thought, richly deserves special mention for his excellent suggestions. I am also indebted to him for providing me with prepublication copies of his forthcoming books: *A Process Philosophy of Religion* and *Religion and Scientific Naturalism: Overcoming the Conflicts*, as well as his paper "Pantemporalism and Panexperientialism."

I would also like to thank my colleagues in the Physics Department: Andreas Albrecht, Steven Carlip, Glen Erickson, and Joe Kiskis for helpful discussions.

I am most grateful to my son, Eric Jungerman, for drawing many of the figures, and to Gera Hasse for her reproduction of the photographs.

Throughout the preparation of this book, my wife, Nancy, has been not only a patient critic, but also a source of loving support. For that I have been especially blessed.

*Introduction to SUNY series in Constructive Postmodern Thought**

The rapid spread of the term *postmodern* in recent years witnesses to a growing dissatisfaction with modernity and to an increasing sense that the modern age not only had a beginning but can have an end as well. Whereas the word *modern* was almost always used until quite recently as a word of praise and as a synonym for *contemporary*, a growing sense is now evidenced that we can and should leave modernity behind—in fact, that we must if we are to avoid destroying ourselves and most of the life on our planet.

Modernity, rather than being regarded as the norm for human society toward which all history has been aiming and into which all societies should be ushered—forcibly if necessary—is instead increasingly seen as an aberration. A new respect for the wisdom of traditional societies is growing as we realize that they have endured for thousands of years and that, by contrast, the existence of modern civilization for even another century seems doubtful. Likewise, *modernism* as a worldview is less and less seen as the Final Truth, in comparison with which all divergent worldviews are automatically regarded as "superstitious." The modern worldview is increasingly relativized to the status of one among many, useful for some purposes, inadequate for others.

---

*The present version of this introduction is slightly different from the first version, which was contained in the volumes that appeared prior to 2000. My thanks to Catherine Keller and Edward Carlos Munn for helpful suggestions.

Although there have been antimodern movements before, beginning perhaps near the outset of the nineteenth century with the Romanticists and the Luddites, the rapidity with which the term *postmodern* has become widespread in our time suggests that the antimodern sentiment is more extensive and intense than before, and also that it includes the sense that modernity can be successfully overcome only by going beyond it, not by attempting to return to a premodern form of existence. Insofar as a common element is found in the various ways in which the term is used, *postmodernism* refers to a diffuse sentiment rather than to any common set of doctrines—the sentiment that humanity can and must go beyond the modern.

Beyond connoting this sentiment, the term *postmodern* is used in a confusing variety of ways, some of them contradictory to others. In artistic and literary circles, for example, postmodernism shares in this general sentiment but also involves a specific reaction against "modernism" in the narrow sense of a movement in artistic-literary circles in the late nineteenth and early twentieth centuries. Postmodern architecture is very different from postmodern literary criticism. In some circles, the term *postmodern* is used in reference to that potpourri of ideas and systems sometimes called *new age metaphysics*, although many of these ideas and systems are more premodern than postmodern. Even in philosophical and theological circles, the term *postmodern* refers to two quite different positions, one of which is reflected in this series. Each position seeks to transcend both *modernism*, in the sense of the worldview that has developed out of the seventeenth-century Galilean-Cartesian-Baconian-Newtonian science, and *modernity*, in the sense of the world order that both conditioned and was conditioned by this worldview. But the two positions seek to transcend the modern in different ways.

Closely related to literary-artistic postmodernism is a philosophical postmodernism inspired variously by physicalism, Ludwig Wittgenstein, Martin Heidegger, a cluster of French thinkers—including Jacques Derrida, Michel Foucault, Gilles Deleuze, and Julia Kristeva—and certain features of American pragmatism.* By

---

*The fact that the thinkers and movements named here are said to have inspired the deconstructive type of postmodernism should not be taken, of course, to imply that they have nothing in common with constructive postmodernists. For example, Wittgenstein, Heidegger, Derrida, and Deleuze share many points and concerns with Alfred North Whitehead, the chief inspiration behind the present series. Furthermore, the actual positions of

the use of terms that arise out of particular segments of this movement, it can be called *deconstructive, relativistic*, or *eliminative* postmodernism. It overcomes the modern worldview through an antiworldview, deconstructing or even entirely eliminating various concepts that have generally been thought necessary for a worldview, such as self, purpose, meaning, a real world, givenness, reason, truth as correspondence, universally valid norms, and divinity. While motivated by ethical and emancipatory concerns, this type of postmodern thought tends to issue in relativism. Indeed, it seems to many thinkers to imply nihilism.* It could, paradoxically, also be called *ultramodernism*, in that its eliminations result from carrying certain modern premises—such as the sensationist doctrine of perception, the mechanistic doctrine of nature, and the resulting denial of divine presence in the world—to their logical conclusions. Some critics see its deconstructions or eliminations as leading to self-referential inconsistencies, such as "performative self-contradictions" between what is said and what is presupposed in the saying.

The postmodernism of this series can, by contrast, be called *revisionary, constructive*, or—perhaps best—*reconstructive*. It seeks to overcome the modern worldview not by eliminating the possibility of worldviews (or "metanarratives") as such, but by constructing a postmodern worldview through a revision of modern premises and traditional concepts in the light of inescapable presuppositions of

---

the founders of pragmatism, especially William James and Charles Peirce, are much closer to Whitehead's philosophical position—see the volume in this series entitled *The Founders of Constructive Postmodern Philosophy: Peirce, James, Bergson, Whitehead, and Hartshorne*—than they are to Richard Rorty's so-called neopragmatism, which reflects many ideas from Rorty's explicitly physicalistic period.

*Peter Dews says that although Derrida's early work was "driven by profound ethical impulses," its insistence that no concepts were immune to deconstruction "drove its own ethical presuppositions into a penumbra of inarticulacy" (*The Limits of Disenchantment: Essays on Contemporary European Culture* [London, New York: Verso, 1995], 5). In his more recent thought, Derrida has declared an "emancipatory promise" and an "idea of justice" to be "irreducible to any deconstruction." Although this "ethical turn" in deconstruction implies its pulling back from a completely disenchanted universe, it also, Dews points out (6–7), implies the need to renounce "the unconditionality of its own earlier dismantling of the unconditional."

our various modes of practice. That is, it agrees with deconstructive postmodernists that a massive deconstruction of many received concepts is needed. But its deconstructive moment, carried out for the sake of the presuppositions of practice, does not result in self-referential inconsistency. It also is not so totalizing as to prevent reconstruction. The reconstruction carried out by this type of postmodernism involves a new unity of scientific, ethical, aesthetic, and religious intuitions (whereas post-structuralists tend to reject all such unitive projects as "totalizing modern metanarratives"). While critical of many ideas often associated with modern science, it rejects not science as such but only that *scientism* in which the data of the modern natural sciences alone are allowed to contribute to the construction of our public worldview.

The reconstructive activity of this type of postmodern thought is not limited to a revised worldview. It is equally concerned with a postmodern *world* that will both support and be supported by the new worldview. A postmodern world will involve postmodern persons, with a postmodern spirituality, on the one hand, and a postmodern society, ultimately a postmodern global order, on the other. Going beyond the modern world will involve transcending its individualism, anthropocentrism, patriarchy, economism, consumerism, nationalism, and militarism. Reconstructive postmodern thought provides support for the ethnic, ecological, feminist, peace, and other emancipatory movements of our time, while stressing that the inclusive emancipation must be from the destructive features of modernity itself. However, the term *postmodern*, by contrast with *premodern*, is here meant to emphasize that the modern world has produced unparalleled advances, as Critical Theorists have emphasized, which must not be devalued in a general revulsion against modernity's negative features.

From the point of view of deconstructive postmodernists, this reconstructive postmodernism will seem hopelessly wedded to outdated concepts, because it wishes to salvage a positive meaning not only for the notions of selfhood, historical meaning, reason, and truth as correspondence, which were central to modernity, but also for notions of divinity, cosmic meaning, and an enchanted nature, which were central to premodern modes of thought. From the point of view of its advocates, however, this revisionary postmodernism is not only more adequate to our experience but also more genuinely postmodern. It does not simply carry the premises of modernity through to their logical conclusions, but criticizes and revises those premises. By virtue of its return to organicism and its acceptance of nonsensory perception, it opens itself to the recovery of truths and values from various forms of premodern thought and practice that had been dogmatically rejected, or at least restricted to "practice," by modern thought. This reconstructive postmodernism involves a creative synthesis of modern and premodern truths and values.

This series does not seek to create a movement so much as to help shape and support an already existing movement convinced that modernity can and must be transcended. But in light of the fact that those antimodern movements that arose in the past failed to deflect or even retard the onslaught of modernity, what reasons are there for expecting the current movement to be more successful? First, the previous antimodern movements were primarily calls to return to a premodern form of life and thought rather than calls to advance, and the human spirit does not rally to calls to turn back. Second, the previous antimodern movements either rejected modern science, reduced it to a description of mere appearances, or assumed its adequacy in principle. They could, therefore, base their calls only on the negative social and spiritual effects of modernity. The current movement draws on natural science itself as a witness against the adequacy of the modern worldview. In the third place, the present movement has even more evidence than did previous movements of the ways in which modernity and its worldview *are* socially and spiritually destructive. The fourth and probably most decisive difference is that the present movement is based on the awareness that *the continuation of modernity threatens the very survival of life on our planet*. This awareness, combined with the growing knowledge of the interdependence of the modern worldview with modernity's militarism, nuclearism, patriarchy, global apartheid, and ecological devastation, is providing an unprecedented impetus for people to see the evidence for a postmodern worldview and to envisage postmodern ways of relating to each other, the rest of nature, and the cosmos as a whole. For these reasons, the failure of the previous antimodern movements says little about the possible success of the current movement.

Advocates of this movement do not hold the naively utopian belief that the success of this movement would bring about a global society of universal and lasting peace, harmony and happiness, in which all spiritual problems, social conflicts, ecological destruction, and hard choices would vanish. There is, after all, surely a deep truth in the testimony of the world's religions to the presence of a transcultural proclivity to evil deep within the human heart, which no new paradigm, combined with a new economic order, new child-

CHAPTER 1

Process Philosophy

Introduction

Rome. The year is 1600. Imagine yourself as part of a crowd that has gathered in the Plaza San Marco to witness the burning of a heretic at the stake. An upstart astronomer, Giordano Bruno, has made the ridiculous claim that Earth goes around the Sun. Everyone you know agrees that Earth is the center of the universe and that the Sun and the planets go around Earth. This trouble-maker has absolutely no respect for church authority and for a tradition going back to Greek times. Additionally he espouses doctrines of pantheism instead of the one God that we all profess. He must be publicly silenced.

Such was the power of the church in matters of science and theology when observations produced the Copernican model of the solar system. It was fortunate for Copernicus that he died only a few months after revealing his model. For while it is true that theological disputes among Christians took thousands of lives, whereas church disputes with scientists were relatively rare, nevertheless scientists soon learned that it was better to quietly ignore the church and to go their own way.

Another Italian, Galileo Galilei, made his own observations with his new telescope and came to the same conclusions as Bruno and Copernicus. Papal authority forced him to recant publicly—or else. Up to these times the church had been the custodian of science, preserving the teachings of Plato over the centuries, and, borrowing from the Moslems, promulgating the teachings of Aristotle as well. The church offered a complete worldview, but now science was beginning its own enterprise.

In the following centuries the authority of organized religion in scientific matters gradually eroded. The Darwinian theory of evolution in the nineteenth century was perhaps the final blow to such authority. Many modern theologians in the mainstream defer to the authority of science in matters of historical fact and the physical world. They have developed scriptural interpretations that recognize the metaphoric and poetic nature of much biblical language. In these interpretations, the creative activity of the divine works through the physical laws of nature rather than contravening them with supernatural powers.

The historic separation of religion and science has had the unfortunate result that now science and technology are generally pursued without moral guidance and traditional religion in the West has become to some extent irrelevant. Process philosophy and the theology derived from it offer assistance in reconciling these important human activities.

The new perspectives offered by science, particularly in the nineteenth and twentieth centuries, influenced many thinkers. Among them was Alfred North Whitehead, who developed a metaphysics in the 1920s and 1930s now termed *process philosophy*. His ideas were fully inclusive and presented as a speculative philosophy and as a descriptive metaphysics. He was aware of the fact that this philosophy would itself be part of an ongoing process and would be evolving. Subsequently, three generations of philosophers and theologians have elaborated a variety of ideas and clarifications in both Christian and non-Christian frameworks based on his seminal work.

Metaphysics ("beyond physics") is concerned with the analysis of experience in the broadest and most fundamental sense. This includes not only our direct experience, but also our interpretation of it: What does it all mean? Does the universe have a purpose? We can only speculate on the answers, and our knowledge is surely superficial and incomplete. Whitehead argues that metaphysics and philosophy permit humanity to cultivate its deeper intuitions: "such as it is, metaphysical understanding guides imagination and justifies purpose. Apart from metaphysical speculation, there can be no civilization."[1]

According to the process theologian David Ray Griffin, in our present day the integration of science and religion is important not only for the vision of harmony it presents, but even more importantly the lack of such integration may imperil the future of civilization. He believes, with Whitehead, that process philosophy can be of service in this regard.[2]

Process philosophy is remarkably compatible with what we have learned about matter and about the evolution of the cosmos. It invites us to look for patterns of interdependence and connection and for the becomingness of events rather than for assemblages of objects. Its task is ambitious: to try to unify all we know about the universe in a consistent system. As we shall see, in Whitehead's formulation of process philosophy a concept of divinity is considered necessary to construct a consistent metaphysics. Thus, process thought forms a natural avenue for the integration of science and religion. With process thought we have the framework for a new myth appropriate to our age to guide our religious quest.

Whitehead introduced process philosophy into the twentieth century with two seminal books: *Science and the Modern World* and *Process and Reality.*[3] These publications signaled his remarkable second career as a philosopher at Harvard University after a distinguished career in mathematics at Cambridge University, where he coauthored with Bertrand Russell the landmark publication *Principia Mathematica*; he was also a philosopher at the University of London.

In his process philosophy, Whitehead attempted to produce a metaphysics that would encompass all that was known, including the new ideas of twentieth-century physics. He greatly influenced a generation of students at Harvard, most significantly his postdoctoral student Charles Hartshorne, who became a philosophy professor at the University of Chicago. Later, a number of graduates of the University of Chicago Divinity School used the ideas of process philosophy as the foundation of their theologies.

This chapter explains the principal ideas of process philosophy. In succeeding chapters we shall see in particular how process thought illuminates and is in full accord with our experience of modern physics and cosmology—from the quantum world of elementary particles, to the human scale, and finally to our understanding of the cosmos and its creation.

Principal Ideas

The main ideas of process philosophy are as follows:

- Events, understood as actual occasions, are primary, not substances.
- Each event is connected to earlier events.

- Its goals are to maximize creativity and intensity of experience, considered broadly.
- Events have an active selection among alternatives.
- Body and mind are interconnected.

Let's now consider each of these ideas in turn.

Events, Understood as Actual Occasions, are Primary, not Substances

Activity and transition are more important in process philosophy than permanence and substance—which is indeed in accord with our most recent scientific ideas about the nature of matter. As we shall see in later chapters, substances that appear to our senses to be solid can be regarded on another level to consist mostly of empty space filled with a continual exchange of virtual particles (photons and gluons) dancing among point masses (electrons and quarks). Thus, from the scientific perspective, permanence and substance are not what they seem to our human senses.[4] In this modern picture, substance is much better described as a series of events.

Process thought views events, not substance, as primary. According to Whitehead, "The simple notion of an enduring substance sustaining enduring qualities expresses a useful abstract for many purposes of life. But whenever we try to use it as a fundamental statement of the nature of things it proves itself mistaken."[5] The idea of inert matter as considered throughout the seventeenth to nineteenth centuries, which is still a pervasive idea, gives us no possible basis for explaining interrelationships—especially those relations conceived in physics as "forces."

Fundamental to process philosophy are *events*, understood as *actual occasions*. In the process view, the fully actual entities are not things that endure throughout time, but momentary events. Actual entities are, thereby, called actual occasions. Such events take place during a short time interval, a fraction of a second, at a particular place. Thus, actual occasions occur in space and time, space-time. We shall discuss space-time in more detail in the next chapter.

Actual occasions occur at different levels, such as at the level of atoms and at the level of human experience. An enduring entity composed of actual occasions could be an atom or an organism, such as a human being. At the most elementary level, electrons and quarks can be understood as a series of actual occasions. For

Whitehead, a moving electron has a different identity at every instant because its position has changed. Its trajectory is a series of events. Whitehead calls this a serially ordered *society of actual occasions*. A human being is a very complex society of events, the dominant member of which can enjoy emergent, unitary consciousness. A human being in process terms is described as a *complex spatiotemporal society of events.*

A distinctive feature of this view is that every actual occasion is to be understood as an *occasion of experience*. We humans certainly have "occasions of experience" and it is reasonable to assume that other animals have them as well. The extension of the idea of experience to lower levels of organization is supported, for example, by research on bacteria, showing that they have a primitive form of memory.[6] Process philosophy assumes that at least some rudimentary form of experience is universal, existing in all actual entities. In this book we shall see evidence for this in the physical world.

The universality of experience has been termed *panexperientialism* by David Ray Griffin. In a recent book, *Religion and Scientific Naturalism: Overcoming the Conflicts,* Griffin says, in opposition to dualism:

> Panexperientialism is based upon the . . . assumption that we can and should think about the units comprising the physical world by analogy with our own experience, which we know from within. The supposition, in other words, is that the *apparent* difference in kind between our experience, or our "mind," and the entities comprising our bodies is an illusion, resulting from the fact that we know them in two different ways: we know our minds from within, by identity, whereas in sensory perception of our bodies we know them from without. Once we realize this, there is no reason to assume them really to be different be in kind.[7]

The viewpoint that events and their relationships are primary is much more in accord with modern physics than with the idea of inert matter. Ivor Leclerc, a philosopher, makes this point in *The Philosophy of Nature:*

> [I]n Whitehead's theory, by reason of the acting being a relating, the whole determines the constituents, and the constituents, by their acting determine the whole. In this

> theory, therefore, the character of the whole "arises from" or "*emerges*" from the constituents, and by virtue of the interrelatedness of their acting this character is not a mere sum of the characters of the constituents. Further, it is by virtue of this character's being mutually shared by the constituents in their acting that the particular character, and the character of the whole, is maintained. It is this which constitutes the "bond" between the constituents, the "force" which holds them together in that particular whole."[8]

We shall see in chapter 4 that in the hydrogen atom the electron probability distribution assumes a particular form that is characteristic of the electron in this environment, and in turn that this distribution forms a bond holding the electron to the atomic nucleus. Again in chapter 5 we shall see that particular quarks are needed to make a proton. These quark constituents are in special interconnections via gluons to bind themselves together to form a proton, the whole.

We experience the weather, other people, and ourselves, which shows that our experiences are not self-contained substances, but processes involving relationships. For Descartes, and for many later philosophers, substances and their qualities were fundamental. Relationships were secondary and even difficult to include in the philosophical systems. This notion of *substantiality* illustrates what Whitehead termed *the fallacy of misplaced concreteness.*

The philosophical idea of substantiality, which has no place for relationships, helped to separate religion from science. Without relatedness to religious questions, science became amoral. It also separated knowledge from emotion and fact from value, for there is little room for emotion and feeling in a world of disconnected objects. It also has led to excessive individualism at the expense of community with profound effects on economics and ecology. We shall consider these effects briefly in chapter 9.

Interdependence and relationships are also fundamental to Buddhist thought. They are examples of several areas of compatibility of Buddhism and process philosophy. The Buddhist doctrine of *pratitya-samutpada*, which means "dependent origination" or "conditioned genesis," asserts that all things *are* by their participation in other things. For the doctrine of conditioned genesis nothing in the world is absolute. Everything is conditioned, relative, and interdependent. This is sometimes called the Buddhist theory of relativity.[9]

In common with process philosophy, Buddhists reject the idea of substantiality (*anicca*). Buddhist meditation practice is not a quest for being, a substance that grounds all things, but a search for emptiness, or perfect peace. Buddhists also reject the idea of a permanent self (*anatta*) since all things are changing, including ourselves. Change is fundamental to process thought.

Each Event is Connected to Earlier Events

Reality is a dynamic organism rather than a machine. Each level of organization—atoms, cells, organs, organisms, and communities—affects all the other levels in a complex *web of interactions* or *connections*. This is Whitehead's "principle of relativity," which has some similarity to the Buddhist one that was just referred to.

Each occasion of experience integrates and uses information in its own incorporation of past particles, fields, and possibilities to produce itself. Each such event Whitehead says, *prehends* the previous events. "Prehends" has an active meaning: "to grasp, or incorporate." It implies that the event is an action—gathering data from the past. This idea challenges the usual notion of cause and effect wherein the affected object is passive. Here not only the causing event but also the affected event takes an active role. *Prehension* describes the connection between past and present events, no matter how elementary. The present event, which is partially self-determining, makes a creative selection among data from all past events and from alternative future possibilities, goals, or aims. Such aims are only potentialities until the selection actualizes them.

When we make a decision at the human level, we incorporate, directly or indirectly, all past experiences, for example, suppose I consider going to a trade school or graduate school. I take into account my past experience, teachers, and what I have studied. Goals are important, too: What is important to me as my life's work? Should I attend now or wait? I also consider all the possible schools. Finally I make a decision to go to a particular school and study a particular subject. I have taken all of this into account, albeit often unconsciously, in making a choice. Once I make the decision, it becomes a fact (objectified) on which I shall base future decisions, such as when I will leave and where I will live. When we have made a decision, this event is actualized and becomes an object that is available for the experience of future entities.

Each occasion gathers information from all previous occasions. At some level we are aware of the mysterious "otherness" of our fellow creatures and take them into ourselves, informing our own actuality—not just in appreciation, but rather in recognizing that the "other" is in some sense constitutes ourselves. We are all inextricably linked in the matrix of creation. We cannot be separate.

As the biologist Conrad H. Waddington observed: "Every event, for Whitehead, contains some reference to every other event in the universe. Every time-extended occurrence to which we give a name, every stone, or table, or person is, as it were, a knot in an indefinitely complex four-dimensional network of relations."[10] Thus, in the process view, everything is interconnected through relationships. In this philosophy there is a place not only to include scientific truth, but also for human perception of spiritual reality.

Interconnectedness, for example, has long been a precept of Native American thought. Jack Forbes, a professor of Native American Studies, expresses its philosophy in his poem "Kinship Is the Basic Principle of Philosophy," the first few lines of which are as follows:

The Thunder-beings are alive:
  grandfathers!
The Earth is alive:
  mother and grandmother!
The trees are alive:
  grandfathers, grandmothers!
The rocks are alive:
  relations of all!
The birds of the air
  the fishes of the sea
  the animals that run
  the smallest bugs
  we are related!

For hundreds of years
  certainly for thousands
Our Native elders
  have taught us
"All My Relations"
  means all living things
  and the entire Universe
"All Our Relations"
  they have said
  time and time again.
. . . .[11]

In the process view, reality is a myriad of serial events somewhat like a motion-picture film, in which the action proceeds from one frame to another in a fraction of a second. One difference in process thought is that each frame is itself a process in which there is a pause for the integration of past events and goals. It would be more in conformity to Whitehead's thinking to view each "frame," or occasion, as actively incorporating, "prehending," all the preceding frames, not only of the movie to which it belongs, but also of all the other movies. In any case, at the moment of decision novelty may be created—there may be something new in the universe.

Therefore, according to process philosophy, the world is a web of interrelated events, a network of mutual influences. The world acts as an interpenetrating field that extends throughout all space, in contrast to the idea of localized self-contained particles. We shall meet this idea of interconnection in quantum mechanics, in particle physics, in complex systems, and in the formation of the universe.

There are times when we can feel intuitively such fundamental interconnection. For me such a moment occurred while backpacking in the Trinity Alps of California. We were camping at an alpine lake with granite cliffs soaring a thousand feet above us. All was quiet except for the brush of a slight breeze in the pines and firs. I experienced an overwhelming feeling of being connected intimately with everything around me. The ferns, the trees, even the rocks, seemed to be part of me. We were all one. A great feeling of contentment came over me. It seemed sublime and natural.

The message was: This is just the way it is. Such spiritual experience is not in the logical methodology of science, yet it was very real to me. Process philosophy acknowledges not just our rational knowledge of the world, but gives feeling and intuition a prominent place as well. Those of the Quaker persuasion might say that such experiences may come when one has a "prepared mind." Mine was prepared in the sense that I had placed myself in a place of pristine beauty removed from the frantic pace of our modern life.

Its Goals are to Maximize Creativity and Intensity of Experience, Considered Broadly

My backpacking experience might be described in more philosophical language: A society of occasions (me) has a certain *intensity of experience*, or *satisfaction*. Every occasion, and every serial-ordered society of occasions, has an intrinsic value, an inner reality for itself. For Whitehead's organic philosophy this experience is "the self-enjoyment of being one among many, and being one arising out of the composition of many."[12]

Since occasions of experience are *acting* entities that incorporate their own past experiences and those of other occasions, new wholes can *emerge* as each occasion manifests its relation to the others. This mutual interrelating may produce a defining characteristic of a new whole and in turn the acting entities are themselves constituted in relation to the whole. This is not possible if the occasions of experience, or events, are changeless and noninteracting, which is the view taken when substances are assumed to be the basic entities.

Cells, for example, may unite to form an organ, such as a human heart. As a result they constitute themselves, emerge, into a new whole. The heart has capabilities that the individual cells did not have. It has unique defining characteristics, and the new organism has an enhanced intensity of experience.

Events Have an Active Selection among Alternatives

Whitehead assumed that *mentality*, at least in some slight degree, is present in every occasion of experience or event. Here "mentality" is not used in the usual sense as only an attribute of humanity. Although human beings are unique by virtue of our very high-level form of mentality, Whitehead regarded a primitive form of mentality to be universal—a bold assertion of his world vision. This is an example of his search for consistency in his metaphysics. He assumes that there is an elementary mentality in every actual occasion that permits it to make an active selection among alternatives.

Figure 1.1 shows schematically the formation of an event, in the sense of an occasion of experience. The event begins with a prehension, a grasping, of previous events, including the prior events of the enduring society to which it belongs, by the physical pole. The physical pole is the part of the event that interacts with the external world. This information passes onto the mental pole. The mental pole is internal to the event. It synthesizes the data coming to the physical pole with the goals of this becoming event. This process is termed a *concrescence*. When the synthesis is accomplished, all the data are simplified in a *satisfaction*, in Whitehead's terminology.

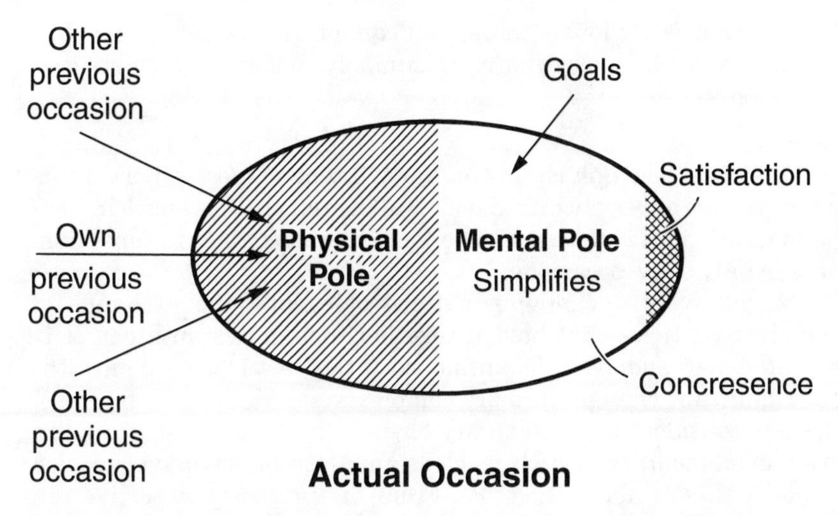

Fig. 1.1. Schematic diagram of an occasion of experience, or actual occasion

The mental pole's creative simplification by selection among alternatives may lead to increased intensity of experience and to novelty. It may lead to a new whole (synthesis) through emergence with other events. On the other hand, a single electron's experience may be viewed as being so dominated by its physical pole that upon concrescence its characteristics are repeated over and over. It becomes an *enduring individual* that retains the same character over a long period of time.

After its satisfaction, an event is said to be *objectively immortal*, as it becomes a datum to be prehended by future events. Note that the event takes place in a finite time, which flows from the past to the future.

Waddington expresses the generalized concept of mentality, which is assumed to be involved in all processes:

> Whitehead was bold enough to take on, face to face, the most difficult of intellectual problems—the fact that each one of us has a conscious experience, whereas in science we try to account for the behavior of things by means of concepts or entities—atoms, waves, fundamental particles, and so on—whose definition does not contain any reference to consciousness. Whitehead argued that this is not good enough: you have either got to have consciousness, or at least something of that general kind, everywhere; or nowhere; and

> since it is obviously in us, and cannot be nowhere, it must therefore be everywhere, presumably mostly in very rudimentary form.[13]

It should be emphasized that for Whitehead most experience is nonconscious experience. In his view all actual entities share in having nonconscious experience, whereas experience rises to consciousness in only a few occasions.

A relatively unchanging pattern of events (the exchange of transient particles that hold it together) dominates an atom. It is repeated over and over. The influence of the past is passed on without significant change. In more complex systems, such as humans, there is considerable opportunity for novelty. As we shall see later, even an elementary particle, such as an electron, may have a choice of paths that is *not predictable*. From the process perspective the electron is making an active choice among possible alternatives. It has an *openness* available to it.

Body and Mind are Interconnected

Process philosophy rejects not only mind-body dualism, but also materialism or idealism. Materialism, while reducing mind to matter, asks us to take our own experience—what we know best in the universe—as secondary. Idealism would have us reduce matter to mind, thereby implying that our own bodies are less than fully real. Dualism permits both mind and matter to exist, but it has little to say about how they interact.

Dualism became ensconced at the center of European thought with René Descartes in the seventeenth century. His philosophy was mechanistic: "Give me matter in motion, and I will construct a universe." One exception did not fit his mechanical model: human thought. A human being was an exquisite machine—but it had a mind attached that must be taken into account. Thus was born the Cartesian doctrine of the dualism of mind and matter.

In process philosophy the reciprocal relation between mind and body is fully natural. In Charles Hartshorne's words: "Cells can influence our human experiences because they have feelings that we can feel. To deal with the influences of human experiences upon cells, one turns this around. *We* have feelings that *cells* can feel."[14] When we are angry or joyful there are measurable biochemical reactions at the cellular level. In process terminology we could have said that our cells "prehend" our angry or joyful feelings.

Dualism usually implies that mind or spirit is superior to matter—a stance that can discourage concern for the present world in favor of a future one. Process philosophy, on the other hand, provides a unifying picture. It views human beings, for example, as different only in complexity from other organisms, or even from so-called inanimate objects—and we are connected to all. This makes it much easier to identify with nature rather than to feel apart from it. Process philosophy, therefore, emphasizes a profoundly inclusive ecology.

Hartshorne makes the point that in separating the body from the mind we "subordinate the concrete to the abstract." What is real, however, is the moment-to-moment becoming of our occasions of experience. Even our personhood or self is an abstraction—in agreement with Buddhist thought, which maintains that the idea of a substantial self is an illusion because we are changing from moment to moment.

Summary

In summary, I gather here the principal ideas of process philosophy:

- *Events*, understood as *actual occasions*, are primary, rather than substances. This gives the world a dynamic quality and promotes relationships. It unites mind and body as an interconnected unity and provides a moral perspective for science.
- An event actively grasps, or *prehends*, all previous events from the universe in varying degrees in its *physical pole.*
- Each event is assumed to have at least an elementary *mentality* that it uses to make a selection among alternatives, taking into account its goals for the future and data from the external world. This process of self-determination is termed a *concrescence* (becoming concrete).
- When the event completes its process of concrescence, it reaches *satisfaction*. After its satisfaction the event becomes *objectively immortal*, being available as a datum for future events.

- Goals are to maximize creativity and intensity of experience, considered broadly.
- A web of prehensions *interconnects* the world. New wholes *emerge* as events interrelate. Mind and body, being composed of the same kinds of entities—occasions of experience—interrelate.
- *The future is unpredictable* for any event because of the alternatives available to it, its *openness*. This is the basis for the *emergence* of *novelty* and *creativity.*

CHAPTER 2

Ideas from the Special and General Theories of Relativity

Los Alamos, New Mexico, mid-July 1945. I was a graduate student working on the development of the atomic bomb with other budding physicists. We knew that a live test, dubbed Trinity, was to take place very soon. Then the rumor spread throughout the scientific community that the test was to take place at 4:30 in the morning on 16 July at a remote site in the Alamogordo desert. Although we had no official reason to be there, my buddies and I were not about to miss this event. After all, we had been working on the project for two years or more, and the test would be the answer to the big question: Would it work?

One of us had a car, so we assembled our sleeping bags and a bit of food and off we went on the afternoon of 15 July southward about two hundred miles to where we thought the test site was. In late afternoon we turned off the main highway from Albuquerque to Las Cruces onto a side road. There was a fence with a Keep Out sign, and as we waited an army patrol jeep passed by inside the reservation. When it was out of sight, we parked; made our way through the fence, and walked about three miles southward away from the highway. There we found a cinder cone that was about two hundred feet high, a hill formed ages ago by a volcanic eruption. It seemed like an excellent place to observe the test, because we could see for miles in every direction.

During the night we were reassured that we were not too close because on a similar cinder cone a half-mile away we could see lights and activity. In the early morning hours there was an intermittent light rain. We awoke early and waited. Four-thirty came and went and after about an hour we began to pack up to try again another day. Somehow, our rumored information must have been wrong.

Suddenly the dawn turned much brighter than midday. I instinctively turned away from the brilliant flash. After a few seconds an awesome sight greeted us: a great roiling orange cloud filled with purple flashes ascended thousands of feet and quickly formed a mushroom-shaped top. We waited for the blast wave to arrive. We could see it stirring up the desert sand as it came. It turned out that we were twenty miles from the explosion so the blast wave was only a slight puff of wind when it reached us. We were very lucky, because if we had been less than that distance we would have risked retinal damage from the flash. Later we found out that over on the other cinder cone the chief scientists of the project were assembled—of course, they had smoked glasses and knew when the "shot," which had been delayed because of the poor weather, was scheduled. We were caught unawares, just as the Japanese in Hiroshima would be three weeks later.

However, we were very aware of the fact that the mushroom cloud contained highly dangerous radioactive fission products and plutonium. We had no idea in which direction the cloud would spread. We half walked, half ran back to the car, covering the three miles in less than an hour.

On the way back to Los Alamos we had time to reflect. We felt a great pity for the Japanese who would soon feel the force of this terrifying weapon. I had an intuitive feeling that the world had somehow changed in a fundamental way—that it would never be the same again. But mixed with these feelings was the elation that the bomb had really worked! It was a tribute to years of labor by thousands, and as a physicist I could appreciate that it was an awesome verification of the special theory of relativity.

This was my first real proof of the idea that mass and energy are really two aspects of one entity, mass energy, and that under the proper conditions one can change into the other. In the explosion I witnessed, an energy was released equivalent to fifteen thousand tons of TNT. Yet the mass converted to energy was less than a gram, an amount of plutonium about the size of an aspirin tablet. Soon, indeed, $E = mc^2$ would become a commonplace in our culture: Mass could be converted to energy and vice versa. From the process perspective we could say that matter itself is part of a process—a process of conversion from energy—and that energy is a process of conversion from mass.

Not only did the special theory of relativity connect mass and energy, but it also brought new insights into the nature of time.

Time is fundamental to process thinking. As I explained in chapter 1, in process thought events are primary. We assess the past, take into account our goals, and make a decision among alternatives that will not only affect our future but will also be used by other events making their decisions. So in an event there is a flow of time from the past, to a decision in the present, and into the future. By showing that space and time are inseparable, the special theory of relativity supports the process view that everything is related to time.

I shall be referring to the ideas of relativity frequently throughout the coming chapters, as they are a fundamental tool in understanding modern physics, which in turn is fundamental to our understanding of the cosmos. This chapter discusses the most important ideas from the special and general theories of relativity and relates those ideas to process thought.

It should be noted that Alfred North Whitehead objected to Einstein's formulation of the general theory of relativity and had his own theory of relativity, which in one form was equivalent to Einstein's special theory. Whitehead objected to the idea that space "curves," which he thought was another example of "misplaced concreteness." On most issues, however, Whitehead agreed with the Einsteinian revolution. As we shall see, Einstein's formulation is very commendable even if it may eventually need to be revised. Moreover, it illustrates process thought, in particular with its emphasis on time and interconnectedness.

Ideas from the Special Theory of Relativity

In the world we live in, we don't notice the effects described by the special theory of relativity. Our world is one where velocities are minimal compared to the velocity of light. Yet in the microworld of atoms and nuclei and in the cosmos itself, the ideas of the special theory are essential for our understanding. For example, we easily assume that clocks tick at the same rate everywhere—that is our experience, but it is wrong. The world of our human senses is narrow and constricted. We don't live in a world of speeds comparable to light speed. There is a lot more out there.

Published by Albert Einstein in 1905, the special theory of relativity remained controversial for several decades. Although Einstein received the Nobel Prize in physics in 1921, it was not for his work on relativity, which was then still too controversial.

The special theory proposed some startling new ideas about the relationship between matter and energy and between space and time. Among these ideas are the following:

- Light always moves at the same speed.
- The laws of physics and the velocity of light are the same in all inertial frames.
- Time is not absolute but is part of a process.
- Space is not absolute but part of a process.
- Simultaneity is relative.
- Mass and energy are interconnected.
- Mass and velocity are interconnected.

Light Always Moves at the Same Speed

Visible light is a form of electromagnetic radiation restricted to a narrow range of wavelengths. Today we are familiar with a much larger spectrum of electromagnetic waves, from long-wavelength radio waves through shorter-wavelength microwaves, infrared (or heat) radiation, visible light, ultraviolet light, and x rays, to gamma rays that originate from atomic nuclei. All move at the same speed.

James Clerk Maxwell published his now famous theory of electromagnetism in the 1870s. It not only explained the known phenomena of electricity and magnetism but also brilliantly predicted the existence and velocity of electromagnetic waves. Maxwell's theory predicts that electromagnetic radiation (of which light is a form) always has the same speed, 186,000 miles per second.

But speed relative to what? Relative to the source that emitted the light? To the fixed stars? The almost universal answer then was that since light was known to be a wave, and since waves must have some medium in which to do their "waving," it must be this medium, whatever it was, which constituted the reference point for the speed of light. No such medium was known, but some of the characteristics that it would possess to propagate light waves could be described, and so its existence was hypothesized. It was named the *ether.*

Nobody doubted that the ether existed. But what would its properties be? Light would move through it while coming to us from the Sun, but in what direction would the ether be moving? In 1887 two American physicists, Albert Michelson and E. W. Morley, set up a delicate experiment to measure the ether velocity.[1] The experiment was conducted at the Case Institute of Technology in

Cleveland. To their complete surprise, Michelson and Morley found no evidence for an ether at all! The experiment was repeated many times in different laboratories and at different times of the year—in case the ether velocity just happened to be zero with respect to Earth at a particular time—but to no avail. Even as late as the 1930s Michelson still thought that he had done something wrong in his measurements, so fixed was the idea of the ether as a medium of propagation for light and for other electromagnetic radiation.

Light just propagates without any medium at all.

The Michelson-Morley experiment was a great puzzle to physicists and they developed several ingenious ideas to explain the negative result. But these ideas were just Band-Aids to enable physicists to keep the idea of the ether. What was needed instead was a bold new approach—the sort of thinking done by a young creative mind. And that was the mind of Einstein.

The Laws of Physics and the Velocity of Light are the Same in All Inertial Frames

Einstein understood that our fundamental ideas of space and time needed to be changed. His two assumptions for special relativity are deceptively simple. Here they are:

- *The "laws" of physics are the same in all inertial frames.*
- The velocity of light is the same in all inertial frames.

Here "laws" is in quotation marks to emphasize the fact that all knowledge is contingent and may change. As explained later, science is a process.

An "inertial frame" is a frame of reference that is moving at a constant velocity and that is not subject to any net outside forces. If you are traveling on a jet in smooth flight, the stewardess can pour some orange juice for you while traveling at six hundred miles per hour just as easily as if the plane were motionless. The plane is moving at a constant velocity and thus is an inertial frame.

The idea that the velocity of light is the same in all inertial frames discarded the idea of an ether and explained the Michelson-Morley experiment. However, to be accepted scientifically, the radical consequences of the special theory of relativity needed experimental verification. This was slow to develop at the turn of the twentieth century but is routine in laboratories today.

Let's see how Einstein's assumption about the velocity of light always being the same affects our concept of time itself.

Time is not Absolute but is Part of a Process

In our normal waking lives we experience time as universal and absolute—it seems the same everywhere. Classical physics, the physics of Newton and Maxwell, is based on this idea. Yet there are subjective experiences in which we have the sense that time is relative. When we are young, time seems to pass very slowly—we can't wait to grow up. Later in life the decades seem to slip by with amazing rapidity. People report that in the instant of a car crash or some other life-threatening event, time appears to run much more slowly, everything happening as if in slow motion. In dreams our experience of time is quite different from waking life: The future and the past are sometimes intermixed and yet time does seem to flow, albeit in a different way. The aborigines of Australia have a highly developed sense of time—dreamtime.

These subjective experiences are similar to the view of time that is fundamental to process thought: It proceeds from event to event, and the events form a web of *connections*. Past events become objects for future events. Now we shall see that relativity also teaches us that time itself is part of a process; it depends on the connection between two societies of events.

The special theory tells us that time depends on space and motion. Since space has three dimensions, we are led to a four-dimensional entity, which is usually called *space-time*. Since space should not have more prominence than time in relativity, we could just as well use "*time-space.*" In fact, Milic Capek argues that since the physics of the special theory of relativity does not eliminate the process idea of becoming, therefore time-space is more appropriate than space-time.[2]

This interconnection of previously considered independent entities agrees with process thought, which emphasizes the fact that the world is formed by a network of connections.

Four dimensions are very difficult to visualize, so let's reduce the number of dimensions. Imagine a car moving along a straight highway. The driver looks at the car clock and notes that it is 3:00 in the afternoon. Let's call this event number 1. After driving a few miles down the road, she looks at the clock again. It's now 3:15. The driver notes that she has gone ten miles. Call this event number 2.

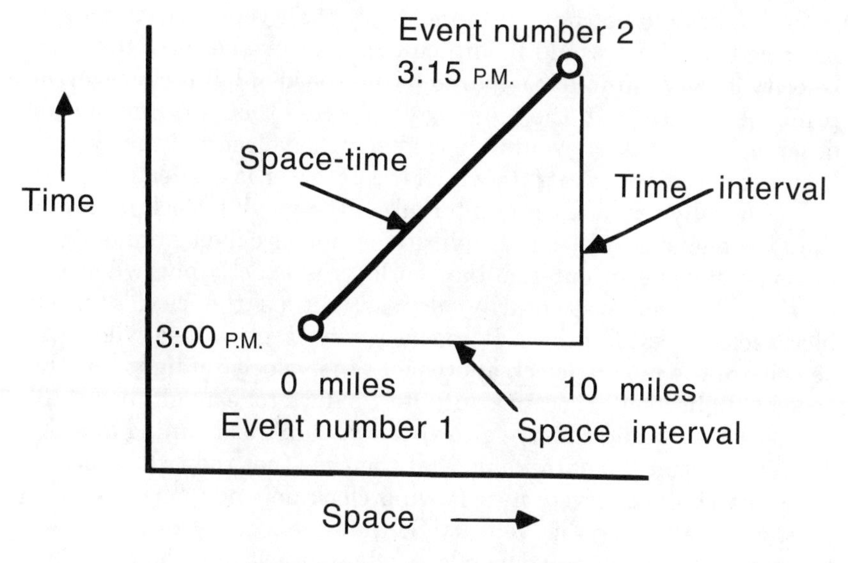

Fig. 2.1. Schematic of space-time

We now draw a diagram representing her trip, shown in figure 2.1. Along the horizontal direction we make two points denoting where she was when event number 1 occurred, and again where she was ten miles down the road. In the vertical direction we place the time as 3:00 to correspond to the first event, and 3:15 for the second event. Note that we are considering a *time interval* here. This is generally true. Relativity is concerned with time intervals, not with absolute time. It is a not a substance-type theory, but can be considered a theory of events.

We have ongoing time as we go up the vertical axis, and increasing distance as we go along the horizontal direction. Now if we connect event number 1 to event number 2 with a straight line, we have a representation of space-time if there is motion in just one direction. The figure also illustrates that events, which are fundamental to process thought, are connected by space-time. But the figure shows only two dimensions, one-dimensional space and time. Space is really three-dimensional, so when we add time, we have four-dimensional space-time, but it isn't easy to draw!

One consequence of the special theory of relativity is the idea that moving clocks run more slowly than clocks at rest. This is called *time dilation*. The faster a clock moves past us, the more slowly it appears to run. In theory a clock in a car moving past you should appear to be running more slowly than your wristwatch. In practice this effect would be difficult to measure because the car's velocity is very minimal compared to the speed of light. However, in principle it is true. In the language of process thought, time for an observer (a society of events) depends on the velocity of the system (another society of events) to which the observer is connected.

Generally the tick (time interval) of the moving clock is slower than the one at rest by a multiplicative number called gamma (γ); γ is very near one in our familiar world; γ is exactly one when the moving clock is momentarily at rest, but as the clock's speed increases, γ's value grows. It reaches very large values when the velocity of the moving clock approaches the velocity of light.[3] At the speed of light itself γ becomes infinite. So if we rode a light beam, all the clocks we would observe elsewhere wouldn't be running at all!

The special theory tells us that the *time* (not the tick, which is the clock *rate*) observed on a moving clock depends on its spatial position as well as on its velocity. So the time a clock tells us is different if it is moving and depends on where it is located as well. This is very different from the understanding afforded by nineteenth-century physics, wherein it was assumed that time was absolute and the same for all frames of reference. In that view, the time indicated by a moving clock would be the same time as indicated by one at rest and would be the same no matter where the clock is.

One of the real-world confirmations that time runs more slowly in a moving system is that muons are observed at sea level. In fact, they are a major component of cosmic rays there. Muons are similar to electrons, except that their mass is 207 times greater. They are produced where the atmosphere begins to be dense, about 12 kilometers above sea level, by very energetic primary particles arriving from interstellar space.

When muons are created in a laboratory, they have a lifetime of only about two millionths of a second. In that brief time, a muon created at the beginning of the thicker, low atmosphere would move toward Earth only 600 meters even if it were traveling at the velocity of light, 3×10^8 meters per second. This is only 1/20th of the distance from where they are created to Earth's surface. So why do we observe them at sea level? They should all decay long before in the atmosphere.

But what if we think of the muons as radioactive clocks? Their lifetime is two microseconds when they are at rest, but if they were created with a high velocity, the radioactive decay would be slowed down. If the decay were twenty times slower than in the laboratory, we could understand why we observe them at sea level. They then would travel on average about 20 × 600 meters, or 12 kilometers, before decaying. If we measure the energy of these muons, we find it is indeed large enough to give them a velocity that produces a value for γ, or time dilation, of 20 or more—a beautiful confirmation of the special theory.

Space is not Absolute but is Part of a Process

Our idea of space is also affected by the assumption that the velocity of light is always the same. Space occupied by a moving object shrinks by just the same factor, γ, by which time is dilated. If we observe a moving object, say a spaceship, its length is apparently contracted in the direction of motion by the same factor, γ, by which its clocks run more slowly. That is, the apparent length would be its length at rest divided by γ. This is called the *Lorentz-Fitzgerald contraction.*

Imagine two spaceships in port. If one takes off and circles the port at half the speed of light, observers in port will see the moving spaceship contracted to about 85 percent of its length at rest. For people on the circling ship, the ship still in port will appear to them contracted to 85 percent of its length! Here the apparent lengths are reduced by dividing the lengths at rest by the factor γ, which is 1.15 for a velocity half the speed of light.

If an object were traveling at infinite speed (really impossible because of the energy required), its clocks would stand still and its length in the direction of motion would be zero.

So we see that space, too, depends on the relationship to the state of motion of another system, another aspect of space-time. The process viewpoint might state it: How we view space depends on relationships, *interconnections*, between two societies of events.

Simultaneity is Relative

Einstein's assumption that the velocity of light is the same in all inertial frames means that whether two events are simultaneous or not depends on who is observing them. Events that are simultaneous for an observer at rest are not for an observer in motion who sees the same events. The process view is that everything depends on *connections*. In that language, simultaneity depends on which society of occasions of experience is observing the same two events. This is illustrated in figure 2.2.

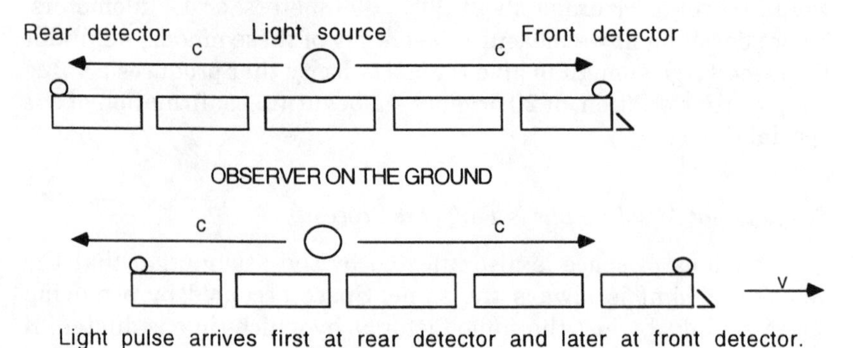

Fig. 2.2. Simultaneity of events and relative motion

We imagine a light signal emitted from the middle of a train that is moving from left to right. To an observer on the train the signal is received simultaneously at the front and rear as depicted by the two arrows moving with velocity c. To someone on the ground watching these events, the velocity of light in both directions *is also the same,* c, but the rear of the train will appear to move into the oncoming light signal and the front of the train will move away from it so that the signal is first received at the rear of the train and later at the front. So the simultaneous events for the observer on the moving train are no longer simultaneous for the observer on the ground.

Even though events that are simultaneous to an observer in motion are not to an observer at rest, the *order* of events is independent of who is doing the observing. In other words, the transformation of space-time produced by the theory of relativity cannot interchange past and future.

Mass and Energy are Interconnected

Fundamental to process thought is the idea that things are connected in a web of relationships. The special theory of relativity not only connects time and space, giving us space-time, but also connects matter and energy, providing us with another transcendent concept: *mass-energy*, a single reality. It indicates that mass can be converted into energy and vice versa.

We shall be using the mass-energy idea a great deal in subsequent chapters: In chapter 3 in the discussion of annihilation quanta, in chapter 4 in relation to spontaneous creation of particle pairs in a vacuum, and in chapter 5 in regard to creating the mass of new particles while investigating the fundamental nature of matter. In chapter 7 we assume that all was radiation at the moment of the Big Bang, but with the passage of an instant in time, matter formed from that energy. As we saw in the introduction to this chapter, mass-energy is the basis for the release of nuclear energy, the energy that makes stars shine. The now famous equation $E = mc^2$ expresses the relationship between mass and energy. Here E is the energy available from a mass m, and c is again the velocity of light. In accord with process thought, we have a new *connection* between mass and energy, whereas in classical physics there is none.

In the case of nuclear fission, the combined mass of the fission fragments from a plutonium or uranium nucleus is less than the mass of the original nucleus. This difference in mass, when multiplied by c^2, gives the energy released per fission, which appears as energy of motion of the fission fragments—kinetic energy. When the moving fragments are stopped by matter, such as in the fuel rods of a nuclear reactor, they create heat. The heat can then be used to make steam, run a turbine, and produce electricity, or, if released violently, produce an atomic bomb.

Mass and Velocity are Interconnected

The special theory of relativity makes another *connection* between two quantities that were considered as separate entities by nineteenth-century physics: mass and velocity. If we observe a moving object, its mass is larger than when the object is at rest, and by just the same factor gamma that we saw in time dilation. So mass and velocity, too, can be considered part of a process of interconnection. I had a very personal experience of this connection when we had just completed construction of a new particle accelerator—a cyclotron.

We were all very tired. The final components of the cyclotron being built by the Physics Department at the University of California, Davis, had been slow to arrive. Then there had been leaks to fix so that we could get an adequate vacuum in the accelerating chamber. And the radio frequency system was balking. The pressure on us was intense, especially for me as director of the project. In just a week the cyclotron was to be dedicated with notables arriving from around the country, including several Nobel Prize winners in physics. It would be so much nicer if the accelerator was functioning instead of having to say: "It is almost working. We're making good progress."

Our cyclotron has a 250-ton magnet that provides a magnetic field six feet in diameter in its accelerating chamber. We were accelerating protons, the nuclei of the hydrogen atom. They were readily formed by introducing hydrogen gas into an electric arc in the center of the accelerating vacuum chamber. Since the protons have a positive electrical charge, an electrode connected to a radio frequency field can give them a small amount of kinetic energy. They then make a half circle in the magnetic field and the radio frequency field gives them another bit of energy. It is a process similar to pushing a child in a swing. Each push sends the child higher and higher. As the protons gain energy, the magnet swings them in ever larger circles.

The time it takes for the proton to make the half circle is always the same, no matter how much energy they have been given, although with more energy they move in larger circles. This is the key to successful acceleration of the proton from the center to the exit radius. The radio frequency needed will be the same for all the acceleration circles.

Well, almost. As the protons gain energy, the special theory of relativity tells us that their mass will increase. To take this into account we carefully tailored the average magnetic field in our cyclotron to increase with radius to track the predicted mass increase. For the protons this increase required the magnetic field to be 7 percent greater at the exit radius.

About 9 P.M. we tried all the systems one more time. Finally everything seemed to be functioning properly. Someone produced a penny. We placed it in a probe that we then inserted into the accelerating region. The penny was on the front of the probe and would be the first to intercept the beam of protons—if they were there!

We ran the cyclotron for a few minutes and then breathlessly withdrew the probe. A Geiger counter was produced. The penny was indeed radioactive! The cyclotron worked! Champagne appeared magically from file cabinets, no doubt sequestered there in hope. It was a great relief for all of us, and a confirmation of the special theory of relativity. The protons really were more massive, 7 percent more than when they started. They had attained a kinetic energy of 67 million electron volts, 7 percent of their rest mass, 938 million electron volts. Gamma had indeed reached 1.07 in Davis for the first time.

The operation of this and of many other accelerators worldwide daily verifies the special theory of relativity. So the amount of matter, mass, itself depends on the *connection* between two different systems of events—a system in which we observe the moving mass and another in which the mass is not moving. If I observed you walking, you would weigh more than if I observed you at rest, but the difference would be very slight. For a person weighing one hundred twenty pounds and walking three miles per hour the weight increase would be about a thousandth of a trillionth of a pound! The difference is so very minimal because walking speed is so slight compared to the speed of light, 186,000 miles per second.

One of the consequences of the increase of mass with velocity is that *nothing with a finite mass can reach the velocity of light*. This is because gamma becomes larger and larger as the velocity approaches the velocity of light. If the velocity reached the velocity of light, gamma (and hence the mass) would be infinite. This imposes a severe restriction on our ability to travel to distant parts of the universe. Even the nearest star is several light years away, and the edge of the universe is about thirteen billion light years.[4] Not only are we restricted by the fact that the maximum velocity is equal to the velocity of light, but also a great deal of energy is required even to approach light speed (except for light itself). Let's look at an example.

If you are a Trekkie, you might want to accelerate the good ship *Enterprise* to, say, half the speed of light to make a trip to the nearest star, Proxima Centuri, about four light years away. It's rather a long trip anyway, eight years one way if we travel at half light speed. The *Enterprise* has four million tons of mass. If we accelerate it to half the velocity of light, then gamma will have reached 1.15. This means that the mass of the *Enterprise* will have increased by 15 percent. How much energy does it take to do that? Since $E = mc^2$, we just have to calculate the energy equivalent of a mass of 15 percent of 4 million tons, or 600,000 tons.

If we convert one gram of mass entirely into energy, it is the equivalent of the explosion of 22,000 tons of TNT. So the energy required to accelerate the *Enterprise* to half the speed of light will be the equivalent of exploding 22,000 tons of TNT per gram multiplied by 6×10^{11} grams, since each ton has about a million grams. So it would take the energy equivalent of about 10^{16} tons of TNT, or ten thousand trillion tons! The nuclear arsenals of all countries contain about twenty billion tons of TNT equivalent. So to put this in perspective, the fuel required for the acceleration to half the speed of light would be the energy equivalent of about five hundred thousand times all the nuclear arsenals of the world. If we did this the *Enterprise* would coast along at half light speed until we reached Proxima Centuri, and then it would take an equal amount of energy to decelerate the *Enterprise* as it reached its destination!

It seems that nature, as revealed by the special theory of relativity, imposes severe limits on our ability to travel even to the nearest star, much less to the far reaches of the universe. According to the general theory of relativity, matter can affect space and time. We therefore have the possibility of warping space sufficiently and cleverly so that the limitations of space travel shown by the special theory of relativity could be overcome. We could imagine that the space between our spaceship and its destination could be compressed. Then our trip would be shortened as much as we might like. Unfortunately the arrangement of mass in the universe to accomplish this would not only be very special, but enormous in magnitude. So it remains a science fiction dream.

Since a mass can't exceed the velocity of light, neither can energy, which is equivalent to mass, or to a signal containing information. Space-time regions that are separated from others such that in the time available a signal from one can't reach the other are not *casually connected.* One of the mysteries of cosmology is why separate and very distant regions in the universe are so similar in composition, yet are so separated that they cannot be causally connected because of the demands of relativity. We shall discuss this again in chapter 7, which is concerned with cosmology.

Ideas from the General Theory of Relativity

Why did Einstein bother to construct a general theory? The answer is that he was not satisfied with the special theory, important as it is, because it is restricted to systems that are moving with constant velocity, which are called inertial systems. The general theory can deal with systems whose velocity is changing—that is, systems that are accelerating, for example, the velocities of planets going around the Sun in their elliptic orbits are continuously changing. Thus, the general theory is a better description of planetary motion than the older Newtonian theory.

Einstein devoted ten years of effort to developing the general theory after formulating the special theory in 1905. It took a long time not only conceptually but also because he had to become familiar with a new area of mathematics in order to describe it. About 1912, when he was a professor at the Technische Hochschule in Zurich, Einstein realized that to adequately describe the general theory, he had to give up the normal (Euclidean) rules of geometry. He came to the conclusion that it was more accurate, instead of thinking of gravitation as warping space and time, to think of the curvature of space-time as the source of gravitation.

Since he knew very little about curved geometry, Einstein appealed for help to Marcel Grossman, an old friend. Grossman taught Einstein the mathematics of non-Euclidean geometry that Johann Gauss and Bernhard Riemann had developed in the nineteenth century. Even though he now had the mathematical tools (Riemannian geometry, which permits a curved space-time), it took three more years before Einstein could fit the pieces together to produce a coherent theory.

Finally, in the fall of 1915 while the world was in the maelstrom of World War I, Einstein, the pacifist, developed ten equations, technically called tensor equations, which link mass-energy and the curvature of space-time. Now a professor of physics at the University of Berlin, he presented these equations to the Prussian Academy of Sciences in November 1915.

The general theory of relativity was on the sidelines of physics until about three decades ago, when the field of astrophysics emerged. New discoveries such as quasars (stellar objects that radiate an immense amount of radiation), black holes, evidence for gravitational waves, as well as the Big Bang theory of the beginning of the universe, all require the general theory for adequate explanation. As a result of these and many other experiments, the general theory of relativity is now firmly established along with the special theory.

The general theory predicts the following:

- Mass-energy produces curvature of space-time, which produces gravity.
- Space-time curvature produces deflection of light.
- Highly concentrated mass-energy produces black holes.
- Time slows in the presence of matter or mass-energy.
- The major axis of the orbits of rotating stellar objects will "precess."
- Accelerating masses produce gravity waves.
- The universe is expanding.

Space-time Curvature Produced by Mass-Energy is the Source of Gravity

Newton's laws are very useful. We used them to send astronauts to the moon and the *Rover* to Mars. However, they are not as accurate as the predictions of the general theory of relativity. They are a good approximation to the way in which the world works when gravity is weak. Then the general theory simplifies into the familiar Newtonian theory but the improved and more general picture provided by the general theory of relativity is much more complex. It also changes our worldview—*We can now consider the fact that gravity is produced by the curvature of space-time and space-time curvature is produced by mass-energy*. If space-time is only slightly curved, we recover familiar Euclidean geometry, Newton's law of gravitation, and the special theory of relativity. Or we can say alternatively that Newtonian theory and the special theory of relativity are included in the general theory and can be obtained from it if the change of velocity is minimal and if gravity is weak.

In the general theory, space-time (or time-space) and mass-energy are interlinked in an ongoing process to produce the gravity we observe—another illustration of the interdependence of concepts previously thought to be independent. This is in accord with Whitehead's principle of relativity, that each level of organization affects the others in a web of *connection*.

Because the general theory of relativity is intimately connected with gravity, in our everyday world the effects it describes are with us at every waking moment. We awake and arise from bed against the force of gravity—our muscles working against Earth's curvature of space-time—and that force is with us throughout the day.

Einstein liked to base his theories on general principles from which he could deduce everything. We saw this previously as a basis for his special theory. His general theory is also based on the constancy of the velocity of light and, in addition on *Einstein's equivalence principle: No experiment can distinguish between a uniform gravitational field and an equivalent uniform acceleration*. By way of illustration we can say that it is impossible to distinguish between being at rest under the influence of gravity and being in space away from gravity if the spaceship is accelerating at the same rate as an object accelerates when dropped on earth. This is illustrated in figure 2.3.

In line with the equivalence principle, the effects of gravity are indistinguishable from the curvature of space-time: Where a large

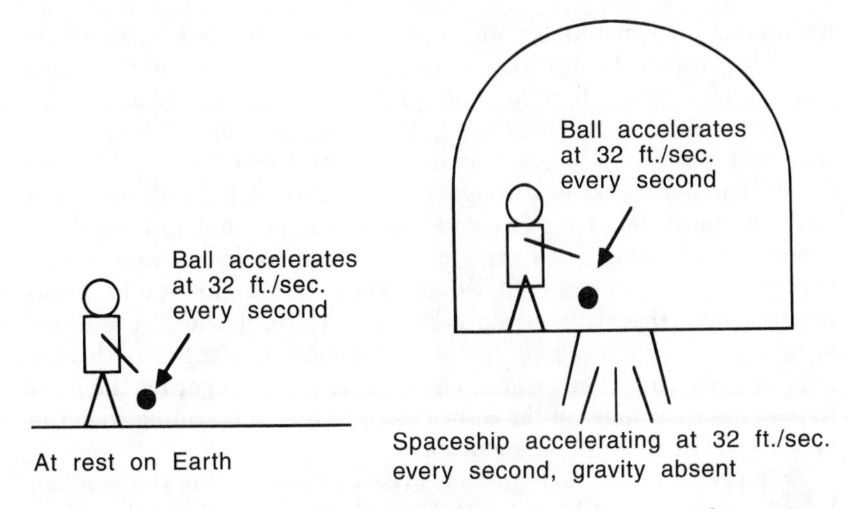

Fig. 2.3. Illustration of Einstein's equivalence principle

mass is present, space-time has a sharper curvature and gravity is stronger. Imagine that there is a stretched horizontal sheet of rubber. If you roll a small ball on its surface, it will go in a straight line. Suppose you take a much heavier ball and place it on the sheet. A depression is formed, gradual far away from the ball and becoming more pronounced as we get near it. Now if we roll the small ball, it will follow a curved path on the rubber sheet. In fact, the small ball can be made to orbit the large one. This is an analogue of, say, Earth's orbiting the Sun by following the curvature in space-time, represented by the rubber sheet. We say space-time rather than just space because mass-energy changes time as well as space.

Light is Deflected by the Curvature of Space-time

One of the first verifications of the general theory of relativity came with a measurement of the deflection of starlight by the Sun. This is possible during a total eclipse when starlight that just grazes the Sun can be observed. Although Einstein's prediction of the deflection of starlight was available when his general theory was published in 1915, World War I prevented any experimental work.

After the war during a total solar eclipse in 1919 in the Southern Hemisphere, two teams of British astronomers verified his prediction. Einstein became an instant international celebrity. It had been argued before that starlight should be bent by the Sun's gravity. However, when Newton's laws are used to calculate the deflection, the result is only one half of the prediction of the general theory of relativity. Observation agreed with Einstein.

When light passes through a gravitational field, it follows a path in space-time that is curved by the mass that produces the gravitational field. If we imagine light coming to us from a star through a gravitational field, we can be fooled. We are used to thinking that light travels in straight lines, so we think the star is located along our line of sight. Instead it is located along a path in a curved *space-time*. Time is also changed. It takes longer for the light to pass near the mass of the Sun. The light is still traveling with the velocity c, but time is going slower.

The bending of starlight that grazes the surface of the sun can be quantitatively explained with Einstein's general theory by calculating the curvature of space-time produced by the Sun's mass. Figure 2.4 shows this effect. If we look through the telescope, the starlight appears to come from along the dashed line, because we assume that light is traveling in a straight line to us. Actually, the light from the star has been bent by the gravitational field of the Sun so that its true position is as depicted along the solid line.

As we shall see in chapter 3, light can be considered as made up of bundles of energy that we call *photons*. Each photon of starlight has a certain energy, E. Since $E = mc^2$, this implies that each photon has an effective mass given by its energy divided by c^2. Then it is reasonable that the mass-energy associated with the photon would be attracted by the Sun's gravity, just as the mass of Earth is attracted by the Sun's gravity so that Earth orbits the Sun. Such a calculation gives only half of the correct result. It takes general rel-

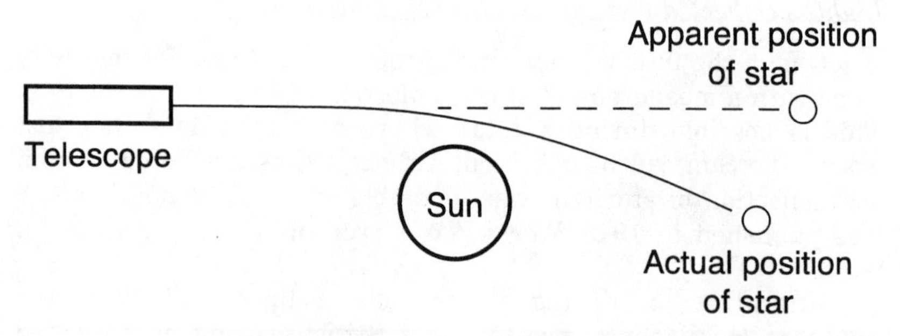

Fig. 2.4. Deflection of starlight by the Sun

ativity to obtain agreement with observation, because the additional gravity produced by the Sun's space-time curvature itself is required.

Another example is the formation of double or even quadruple images of a distant light source by intervening stellar dust that acts as a series of gravitational lenses, thereby producing multiple images. Stellar dust forms the lenses by bending the light in a manner similar to the deflection of starlight by the Sun. Quasars that are eight billion light years distant have been observed to make such images. We think that quasars were formed in the early history of the universe as matter fell into enormous black holes. Some of them had the energy output of a trillion suns! That's why we can see them now, even though the light from them has taken eight billion years to reach us.

This brings us to black holes—a spectacular prediction of the general theory.

Highly Concentrated Mass-Energy Produces Black Holes

Matter in a very condensed state forms black holes. If there is sufficient mass-energy in a small volume, space-time will have so much curvature near it that light or anything else that is moving will follow a curved path that returns to the gravitational source, giving us a *black hole*. A black hole is really not a hole at all, but a region of space-time with so much matter and energy within it that it forms its own universe. If all Earth's mass were concentrated into a sphere the size of a marble, it would form a black hole.

The maximum distance from the black hole that light can go is called the *event horizon*, as illustrated in figure 2.5. It is called the event horizon because we are ignorant of an "event" taking place within a sphere of that radius. This is a microuniverse that is closed to our observation because light can't escape from it. A way of thinking about the event horizon is to remember that a photon, or quantum of light, has an effective mass of E divided by c^2. In trying to leave a black hole, this mass is attracted by the black hole's gravity. So the photon loses more and more energy as it travels outward. By the time it reaches the event horizon, the photon has lost all its energy to the gravitational field and then returns to the hole.

There is increasing evidence that black holes really exist. Since light can't get out of a black hole, we can't see one directly. However, we can calculate the characteristics of radiation, in particular intense x rays, emitted just before matter descends into a black hole.

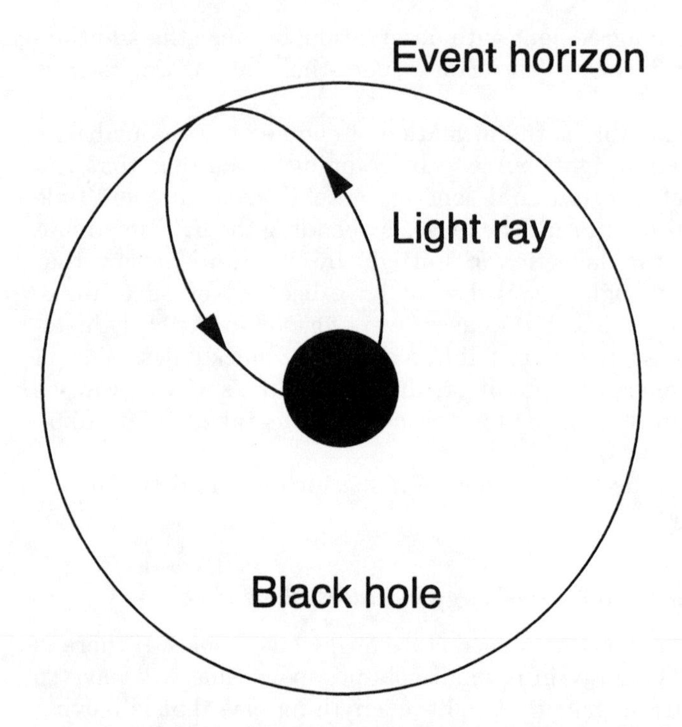

Fig. 2.5. A light ray trying to escape from a black hole

In 1994 the repaired Hubble Telescope was able to measure this radiation and it agreed with the calculation, giving us strong evidence for their existence.

In 1997, observations were reported of stellar motion near the center of our own galaxy. These measurements were done with infrared radiation. We cannot see the central region of our galaxy with visible light because it is hidden by too much dust. The motion of the stars in the central region reveals a compact object of about three million times the mass of the Sun at the galaxy's center. It can only be explained by postulating that it is a black hole.[5]

Black holes were until recently thought to last forever. But in the 1970s Stephen Hawking and his coworkers theorized that some particles could be emitted from black holes, now called Hawking radiation.[6] This radiation is yet unobserved, and would generally be very weak. Their theory developed an ad hoc addition of quantum mechanics to the general theory of relativity.

For a black hole of solar mass to evaporate an appreciable fraction of its mass in Hawking radiation would require 10^{54} times the age of the universe. However, it would require only thirteen billion years, the age of the universe, for a black hole of mass 10^{12} kilograms. This is about the mass of a sphere of Earth's density of about one-half mile in diameter.[7] Thus, it is possible that even black holes evolve and in time evaporate as matter-energy is lost through this radiation. Even black holes are in process.

Time is Slowed Near Mass-Energy

One of the predictions of the general theory is that *time slows down in the presence of mass-energy*. It does so in proportion to the strength of the gravitational field that the mass-energy produces. Again we have an example of the *interconnections* that are emphasized by process thought. Time is not absolute, but depends on connections to mass-energy. Time is affected by a gravitational field, that is, by the presence of mass-energy. As we saw earlier, time also depends on connections to moving systems.

Since the gravitational field of Earth weakens as we go into space, it is possible to test this idea directly, as was done in 1976 by flying an atomic clock on a rocket to an altitude of about six thousand miles and comparing its frequency (tick rate) to that of a similar clock on the ground. The clock on the rocket ran faster, as predicted, and the measured difference agreed with the prediction of the general theory of relativity to within 0.02 percent.

Near the enormous gravity of a black hole, clocks would run very slowly indeed. If we were to observe a light signal emitted by an unfortunate spaceship approaching a black hole, it would get lower and lower in frequency, or longer in wavelength, as the atoms that made the light acted as clocks that were slowing down. One can think of the waves of light getting stretched. The light would get redder, shift to infrared or heat radiation, and finally to radio waves before disappearing altogether. All of this would take a very long time for us, but for someone on the spaceship nothing special would seem to happen at the event horizon.

Planetary Orbits will Precess

Experimental verification of the general theory came immediately with an explanation of a long outstanding problem in astronomy. Mercury's orbit is not fixed, but its major axis rotates slowly with time, a phenomenon known as *precession*. This is a minor but readily measurable effect that had long puzzled astronomers. According to Newtonian physics, Mercury would orbit the sun in the same elliptic path without change. According to observation Mercury precesses, forty-three arc-seconds per century.

If we imagine the gravitational influence of two planets, in the Newtonian theory we would add the contribution of each at a given point to obtain the total gravitational influence of both. In the general theory, the curvatures of space-time produced by the individual planets cannot simply be added to give the combined curvature of both planets. The curved space-time of a given planet contains energy and hence provides further curvature. So curvature of space-time is a source of further curvature or, put another way, in the general theory of relativity, gravity is a source of additional gravity. As a result, the curved space-time around the Sun and around the planets produces more space-time curvature so that the planets do not follow the elliptic orbits predicted by Newton's laws.

Using our rubber sheet analogy, we might visualize the heavy ball representing the Sun placed on the sheet to produce a deformation. We then remove that ball and place another ball that represents the planet Mercury. Now imagine that we place both balls on the sheet. To be in analogy with the general theory, the combined deformation will *not* be the sum of the individual deformations. We might say that the elastic properties of the rubber were changed slightly by the balls so that the deformation will be different if both are there than by simply adding the individual deformations. This corresponds to the curvature of space-time itself producing further curvature.

It has been only recently that the theory was invoked again to explain the phenomenon of orbital precession, but this time concerning a binary pulsar. A *binary pulsar* consists of two neutron stars rotating around a common center. They consist of matter at densities of the atomic nucleus, where a spoonful would weigh a billion tons. They are predominantly of one nuclear species, the neutron. Joseph Taylor and Russell Hulse of Princeton University observed the extraordinary binary pulsar PSR 1913 + 16 for eighteen years.[8] These observations produced enormous precision in the determination of its orbit. In the binary system measured, each neutron star had a mass about one and a half times that of the Sun.

One of the neutron stars in this binary pulsar emits a radio beam pulse of exquisite precision. This pulse that strikes Earth about sixteen times per second. Its precision is as exact as the best atomic clocks. On 14 January 1986, for example, the frequency of the pulsar was 16.940,539,184,253 pulses per second with an error of one in the last digit, or a precision of one part in one hundred thousand billion. As the binary stars rotate about their common center approximately every eight hours, the pulsar's frequency lowers as it moves away from Earth and rises as it approaches Earth (Doppler shift). This frequency is the average value. Figure 2.6 shows this binary pulsar system schematically.

The planet Mercury has a rotation of its major axis, as just noted, at the rate of 43 seconds of arc per century, but PSR 1913 + 16 does it at a rate 30,000 times faster, or four degrees per year. This faster rate enables a much better test of the theory, and it agrees very well.

Accelerating Masses Produce Gravity Waves

Taylor and Hulse's observations also gave us the first evidence of the existence of gravity waves, a fundamental discovery. Gravity waves are ripples in space-time that carry energy away from their source of accelerating masses at the speed of light. They are similar in this respect to electromagnetic waves, which carry energy away from their source of accelerating electric charges, also at the speed of light.

Despite extensive searches in the last few years with special detectors, gravity waves have not yet been detected directly. However, by observing the binary pulsar PSR 1913 + 16, Taylor and Hulse have established definitively the existence of gravitational radiation. Measuring these signals, they were able to demonstrate

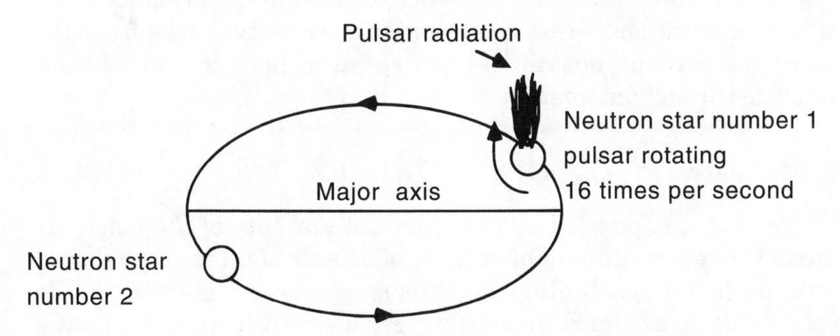

Fig. 2.6. The binary star system PSR 1913 + 16

that the *rotation rate of the binary stars about each other is increasing*, and at precisely the rate predicted by the general theory of relativity, by taking into account the gravitational radiation emitted. In 1993, they received the Nobel Prize in physics for this research.

The two neutron stars are analogous to a giant rotating dumbbell. Each star is accelerating continuously inward as it is constrained in its orbit. The acceleration causes gravitational radiation. As gravitational energy is released from the system, the stars should rotate faster around their common center, that is, the rotation frequency should increase. Taylor and Hulse compared the predicted increase to their accurate observations of the orbital frequency change with time. The result was a stunning success for the general theory of relativity. The ratio of the observed to the calculated frequency change agreed to better than 1 percent. In addition, due to the gravitational radiation, the precession rate of the binary star's major axis itself is growing faster with time. This is also accurately predicted by the general theory.

These measurements are the first clear evidence for gravitational radiation. There are several promising experiments now underway to try to detect this radiation directly. If successful, they will open a new area for investigation of the universe: gravitational astronomy.

Because of gravitational radiation, the binary stars of PSR 1913 + 16 will come closer and closer as they lose energy. Eventually they will combine because of their mutual gravitational attraction and they may produce a black hole, a specific example of the evolving universe. Evolution at every level is one of the basic tenets of process thought. Another fundamental consideration of process thought is interconnection that provides influence from a previous event on a future event. The existence of gravitational radiation is a further illustration of this connectedness. Gravity waves from the binary star expand outward at the speed of light to spread their influence throughout space.

The Universe is Expanding

Newton was puzzled by the apparent stability of the stars. He realized that according to his theory of gravitation the stars should mutually attract each other and coalesce into one giant mass. To explain this he assumed that there were an infinite number of stars so that the gravitational forces on an individual star would all cancel out. There would be no center to collapse toward. But we now know that stars are all moving away from each other: The universe is expanding. Amazingly the general theory of relativity explains this cosmic expansion.

When Einstein produced the general theory in the period 1905–15, the universe was thought to be fixed in volume: There was no evidence for its expansion. Accordingly Einstein added an ad hoc term to his theory to prevent it from predicting an expansion. This term is now called the *cosmological constant*. If it is set to zero, the original theory is recovered.

In the early 1920s the American astronomer Edwin Hubble, working at Mt. Wilson observatory in Southern California, was able to show definitively that the universe is, in fact, expanding. He did this by using Cephid stars, which have a variable brightness. The brightness of these stars, known since the nineteenth century, depends on their period of pulsation. The brighter the Cephids are, the more slowly they pulsate. Cephids are intrinsically bright, typically having luminosities of a few thousand suns. If we observe a Cephid star that is pulsating slowly but is dim, we know that the star is very far away. So the Cephids produce a distance scale. Hubble used this to establish that all stellar objects are moving away from us, and that the farther away they are, the faster they are moving away—just as Einstein's original formulation of the general theory, without the cosmological constant (or with it set to zero), had predicted.

In response to Hubble's evidence, Einstein said that invoking the cosmological constant was the biggest blunder that he had ever made.

If the cosmological constant is zero and if we apply the general theory of relativity to the whole observable universe, it predicts three possibilities for its evolution. Common to all the possibilities is that the universe is expanding, and that the expansion will slow down because of the mutual gravitational attraction of the mass-energy within it. The first possibility, positing a closed universe, suggests that if the total universe has sufficient mass-energy, it could eventually stop and reverse itself into a compression leading to a Big Crunch, that is, the entire universe could collapse into a very small and intensely hot region similar to the Big Bang from which we believe the universe was created. Another possibility posits an open universe. In this case there is insufficient mass-energy to contain the expansion and, although it will slow down, it will never stop expanding. The third possibility, a flat, or Euclidean, universe, is that the universe will slow down in such a fine-tuned way that after an infinite time, it will stop expanding. Recent measurements indicate that space is flat, but there is considerable uncertainty in this conclusion because present data are not precise enough.

According to cosmological data in 1999 the cosmological constant may not be zero after all. They show that the expansion of the universe is now speeding up, apparently due to an antigravity force not yet understood. The additional energy density provided by this force makes space flat. Thus our present knowledge envisions a flat universe that will expand forever (for more detail see chapter 7).

Hubble's discovery of the universe's expansion also showed that it is much larger than previously thought. His discovery, presented to the American Astronomical Society in December 1924, revolutionized our perception of the universe, for example, the Andromeda Nebula was thought to be part of our galaxy, the Milky Way. Hubble observed Cephid stars in the nebula and showed that it is much more distant than stars in the Milky Way. It really is an enormous galaxy about two million light years from Earth. Hubble had discovered the realm of the galaxies. We now know of over 100 billion galaxies, each with about 100 billion stars. All of this is part of a creation story that is only a few decades old.

Let's consider our rubber sheet again. Imagine that it is being stretched uniformly in all directions. If we drew a circle on the sheet, it would become larger as the sheet stretched. This is the way the universe is expanding: Space itself, analogous to the rubber sheet, is expanding. Galaxies and stars seem to move away from us because the space between them and us is expanding. The more space there is between us and them, the faster they recede. Actually, galaxies and stars are held to fixed dimensions by their own gravitational attraction. A better analogy for them on our rubber sheet is to imagine that instead of painting a circle on the sheet we place plastic circles on it. Then as the sheet expands, each circle will find itself farther from the others, but with its own dimensions unchanged.

If we imagine ourselves in one of the galaxies on the rubber sheet, then all the others are moving away from us: The farther they are away, the faster they move. But we would have observed the same thing if we were on a different galaxy. There is no center to the expansion of the universe: All points are equally the center. It is omni-centered. We are all at the center of the universe! Another analogy of the expansion that is three-dimensional, as is actual space, involves raisins in a rising loaf of bread that it is expanding before it is baked. Each raisin finds its neighbors receding from it, and the more distant the neighbor, the faster it recedes.

According to process thought, the divine has been part of the creation of the universe from its beginning and remains immanent in every part of it. Confronted with the scale of the universe, it is easy to feel that an individual may be lost in its immensity. But this is really an assumption. If every point is a center in which the divine is immanent, then all aspects of the universe are equally important. The omni-centered universe is compatible with the process view that all events are in a web of relationships, with each event as centered as all the others.

It is important to know that there is no limit on the recession velocity of space in general relativity. The recession velocity is measured to be about 21.5 kilometers per second for every million light years that a star is distant from us (one light year is the distance light goes in a year, or about 6 trillion miles). This measurement has an experimental error of about 10 percent.

If we ask how far a star has to be for its recession velocity to be equal to the velocity of light, then we divide 21.5 kilometers per second per million light years into 300,000 kilometers per second, the speed of light. That gives us 14,000 million light years, or 14 billion light years. This means that starting from the beginning of our universe light would have traveled 14 billion years to reach us. So that is one estimate of the age of the universe. At the limits of experimental error, this value is consistent with the approximately 13-billion-year figure obtained by other cosmological measurements.

This estimate assumes that the expansion is constant, which is not really the case (see chap. 7). When the recession velocity is equal to the velocity of light, this is the limit of our observable universe. This observable universe is sometimes called the *Hubble sphere*. At greater distances the recession velocity exceeds the velocity of light, so that part of the universe is unobservable to us. So space can expand faster than the light it contains.

Another way of thinking is to imagine time running backward and then to ask how long it would take the space between two galaxies to contract so that they collide. The answer is again 13 billion light years. This result is independent of the initial distance between the galaxies, since if they were initially farther apart, they would have a higher velocity when space contracted (remembering that the expansion velocity is proportional to the distance from any point in the universe).

If we were in another galaxy, then we would also have a limit of the observable universe as 13 billion light years, but it would not have the same stellar constituents. It would have its own Hubble sphere. There are as many Hubble spheres as there are galaxies, some 100 billion! There is no edge to the universe, only an edge to what we can observe. If we imagine a star just beyond the 13-billion-light-year limit of our Hubble sphere, then its light would not be visible. That is because there the velocity of the expansion of space would be greater than the velocity of light.

Science as Process

The discovery of relativity illustrates how science itself is a process. First there is a hypothesis that is invoked to try to explain observations of nature. A good hypothesis is then further elaborated into a theory that not only explains the given observations but is also able to predict the results of new observations. Finally, if a theory has been verified for a sufficient time, its premises or postulates are sometimes given the status of a law of nature. By "laws" of physics we generally mean the equations that describe the behavior of physical systems, ranging from the nucleus of the atom to our everyday world, and to the behavior of stars and galaxies. Newton's law of gravity and his laws of motion are examples. Relativity is still recent and has the tentative status of a theory, but eventually may become the law of relativity.

For science to proceed, a theory must be testable. If it is a good theory, it will also have within it enough basic understanding to make predictions for experiments not yet performed. A superb theory, such as the theory of relativity, not only does all of this, but also gives us a new view of the world and invites us to expand our thinking. The discovery of gravitational radiation is a recent example of the interaction of theory with accurate observation that lies at the heart of science.

Science is also a process in the sense that even the so-called laws themselves are only contingent and sometimes need to be modified as our understanding of the universe increases. Science is continually evolving. Even our most cherished theories are not forever. The laws of Newton are very useful and we do not have to discard them, but we now know that there is a limit to their applicability. The theories of relativity are more general and Newton's laws become a special case, applicable when velocities are much less than the velocity of light, and when the density of mass-energy is minimal.

Thus the old "laws" of nineteenth-century physics about space and time as well as matter and energy are included in, but superseded by, the new ideas of relativity. Science is an ongoing process, in continuous evolution.

One may well ask a theological question: Why do these "laws" exist—not just now, or here on Earth, but at least ten billion years ago and at ten billion light-years distant? Such order is the basis for the evolution of complexity and novelty in the universe, a concept that is basic to process philosophy and theology. For the latter, the order in the universe is the result of the divine "primordial nature" (see chap. 8).

As ideas are introduced throughout this book, it may be helpful to have a sense of the chronological order. They are shown in figure 2.7. The figure also illustrates that science is a process.

Descartes's substantialism was introduced in the latter part of the seventeenth century as was Newton's law of gravitation and the latter's idea that light consisted of particles. In 1800 Thomas Young's double-slit experiment, to be discussed in the next chapter, proved that light had a wave aspect. From 1850 forward, modern physics and cosmology has had an accelerated development. Alfred North Whitehead's book, Process and Reality, which formed the basis for his process philosophy, was originally published in 1929. Only events that are discussed in the present book are shown. Words such as "electron" and "muon" in the figure denote the time when these particles were discovered.

Process Thought and Relativity

As we have seen, time is of the essence of process thought, as the universe goes forward from event to event. Due to special and general theories of relativity we must consider time differently and nonintuitively. In our familiar Newtonian world, time is absolute. It is the same everywhere whether a clock is in motion or not, or whether the clock is in a gravitational field or not. This is in accord with our experience in a world where speeds are very slight compared to the speed of light, and where gravity is relatively weak. But the world looks different when velocities of objects, such as clocks, are comparable to the speed of light, and when gravity is immense, as near a black hole.

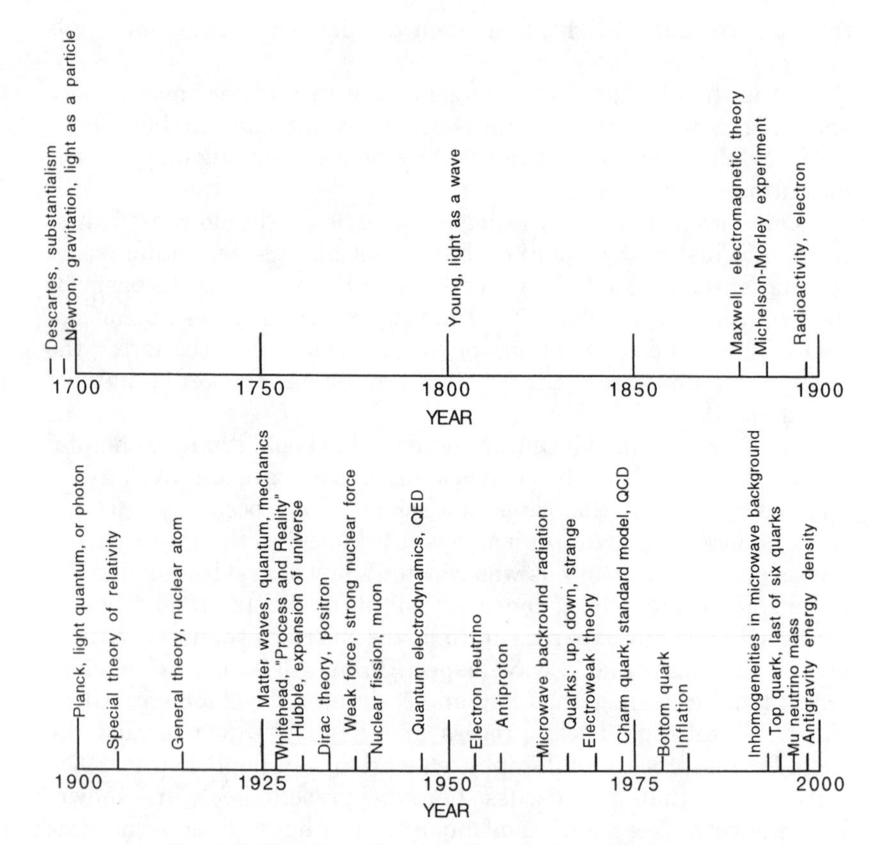

Fig. 2.7. Timeline of events discussed in this book

The special theory of relativity tells us that mass and energy are two forms of the same reality, mass-energy, and predicts that they are in principle interchangeable. Many physical processes demonstrate this, such as the production of pairs of particles from electromagnetic energy and the destruction of electrons and positrons to provide mass-energy in a new form: annihilation radiation (to be discussed in the next chapter). Thus matter itself can be considered as part of a process.

It is a fact of nature that the velocity of light is the same in all systems, whether moving or not, or whether under the influence of gravity or not. Einstein's genius was to recognize this and to use it as a cornerstone of both the special and general theories of relativity. The consequences of this assumption change our ideas not only of time, but space as well. We now understand that space and time, previously thought to be separate, are interconnected. In the new view, space and time are transcended by a unified concept, space-time, which mixes the two together in different ways that depend on the velocity of the observer. The time of an observer at rest depends not only on the time seen on a moving clock, but also depends on where that clock is located in the moving system. A moving object is contracted in the direction of motion. Events that are simultaneous for at observer at rest are no longer simultaneous to one who is moving.

In his general theory of relativity, Einstein recognized the fact that not only is the velocity of light the same for a system at rest and for one that is accelerated, but also that the latter is indistinguishable from a system that is at rest and acted upon by gravity. The theory shows us that the presence of mass-energy slows down time. It also explains gravity as a consequence of the curving of space-time caused by mass-energy. The curvature of space-time itself produces further curvature, or an addition to the gravitational field. Mass-energy tells space-time how to curve, and space-time tells mass-energy how to move.

In the nineteenth century, matter, motion, time, and space were all considered separate realities. Now, in accord with process thought, they are all related and interdependent. The mass-energy, time, and space we observe depend on motion, and space and time change in the presence of mass-energy.

Thus, there is an interdependence of time, space, mass-energy, motion, and gravity. They are all interlinked as part of a process. The general theory also predicts that accelerating masses, such as stars rotating in an orbit, will give off gravitational radiation—a prediction recently verified with great precision.

The observed expansion of the universe, which is predicted by the general theory of relativity, shows us that the universe is indeed not static, but dynamic in its entirety in accord with process philosophy.

The special and general theories of relativity give us a *connection* between concepts that we previously thought to be separate. Such interrelatedness is in accord with process thought. For example, the philosopher of religion and physicist, Ian Barbour discussing the metaphysical implications of the theory of relativity, says that "it shows us a dynamic and interconnected universe. Space and time are inseparable, mass is a form of energy, and gravity and acceleration are indistinguishable. There is an interplay between the dynamics of matter and the form of space, a dialectic between temporal process and spatial geometry. Matter is, if you will, a wrinkle in the elastic matrix of space-time."[9]

In this regard, it could be said that a particle, which concentrates mass-energy in a spatial location, is just a singularity in the space-time metric. We shall consider particles at various times throughout this book, but it is useful to remember this alternative viewpoint. Process thought makes one suspicious of the word "particle," which connoted a changeless inactive substance in nineteenth-century physics. This idea is not in accord with the modern view of matter, as we shall see.

In process philosophy, Whitehead also has a principle of relativity that is compatible with Einsteinian relativity. Whitehead's principle is that each event in the universe influences subsequent events. In process thought, each event is the result of its choice among alternatives that its history partially determines. What we choose as a reference system for the speed of an object also depends on our previous history—that we are here and not somewhere else. A different observer would have a different history and might well determine the speed to be different. All observers are equally correct, and they are interconnected through the relationships of the special theory of relativity. Whitehead was fully informed about the revolutionary nature of Einstein's theories of relativity and notes it in Process and Reality.[10]

The general theory of relativity is essential to our understanding of the evolution of the universe because it predicts its observed expansion. This expansion is fundamental to the current Big Bang theory of its origin. As we shall see (chap. 7), expansion provides a universe that is dynamic and creative, producing amazing novelty. Immense creativity has occurred in the universe since the Big Bang, which initially was just a fireball consisting only of intense radiation. As we study the way the world is, from subatomic particles to the cosmos, we shall find it very compatible with the idea of the primacy of events rather than substance: a basic tenet of process thought. The way in which the universe is constructed, events lead to evolution at every level, even for black holes.

Thus the general theory is key to our understanding of the cosmos and is fundamental to the foundation of a new myth connecting science and religion. In the last few decades, astrophysicists have increasingly employed it not only to understand the expansion of the universe, but also to explain observations of black holes, quasars, orbital precession, and gravitational radiation.

The development of both the special and general theories of relativity demonstrates that science itself is a process. Even such revered ideas as Newton's laws are subject to modification. Science is contingent and ever evolving. No ideas are fixed for all time.

It is sometimes erroneously stated that the special theory of relativity shows that everything is relative—there are no absolutes. But the velocity of light is absolute. It is the same in all frames of reference, even accelerated ones. The laws of physics also are absolute in this sense—they are independent of reference frames.

The theory of relativity, although radically changing our concepts of time, space, energy, matter, and gravity, has a deterministic form in that its predictions are only limited by the available initial and boundary conditions. It also assumes the classical idea that there can be an observer who objectively measures a system without affecting it. Both of these assumptions turn out to be untrue in the physical world. As we shall see in chapters 4 and 6, quantum mechanics and complex systems tell us that the world is really nondeterministic and that the observer is linked to the system observed. In physical systems there is a fundamental uncertainty that makes deterministic predictions either impossible or limited in precision.

CHAPTER 3

The Microworld, Part I: Waves and Particles

What is the world really made of at the microscopic level beyond our human senses? We begin by focusing our attention on light—something we experience in every waking moment, yet full of mystery when we investigate it closely.

In this chapter we look at some of the evidence that light is a wave, and other evidence that it is a particle. In trying to reconcile the two concepts we are forced to consider that light has both aspects—a *wavicle*, a concept fundamental to quantum mechanics, which is our best description of the microworld. That world is closely linked via today's astrophysics to our knowledge of the cosmos and to its origin in the Big Bang. We shall further see that matter, which we ordinarily think of as consisting of particles, also has a wave aspect.

We know from direct experience that light is not a substance, like a table. It is more akin to energy. It originates from a source, goes outward, and is absorbed—a series of events. Throughout this book we shall find increasing evidence that the world is basically composed of events and process rather than static substances. We shall also find that light has both a wave and a particle aspect. They are interconnected in a new transcendent reality, a *connection* that is compatible with process thought.

Evidence that Light is a Wave

Light is a wave. But what is waving? For water waves there is a medium, water, which waves. Individual water droplets go up and down as the wave passes. As we saw in the last chapter, for light the waving occurs in what is effectively a vacuum. What is waving is an electric field accompanied by a magnetic field that waves in synchronism with it. Both fields are at right angles to each other and to their direction of motion.

The electric and magnetic fields rise and fall with distance, or, if we are in a fixed location as they pass by, they rise and fall with time in the same manner. The electric field of the wave is shown in figure 3.1. In the figure the whole wave is moving to the right with the velocity of light, *c*.

For visible light the waving is very fast indeed—several hundred trillion vibrations each second. The distance between waves is much smaller than we can see. For yellow light there would be about a thousand waves in the thickness of a fingernail.

According to James Clerk Maxwell's theory of electromagnetism, published in 1873, when electric charges accelerate, they produce electromagnetic radiation, including visible light. The radiation occurs in different wavelengths depending on the vibratory nature of the source: the longest waves are radio waves, then waves for television transmission and microwaves for your kitchen. Shorter wavelengths are infrared, such as the heat waves we notice from our fireplace, and then to the visible, red to blue. Finally, electromagnetic radiation is found in even shorter wavelengths in the ultraviolet that gives us sunburn, the x rays at our dentist, and the shortest waves of all, the gamma rays that arise from radioactive atoms.

In a triumph for Maxwell's theory, the electromagnetic waves that he predicted were observed by Heinrich Hertz in Germany in

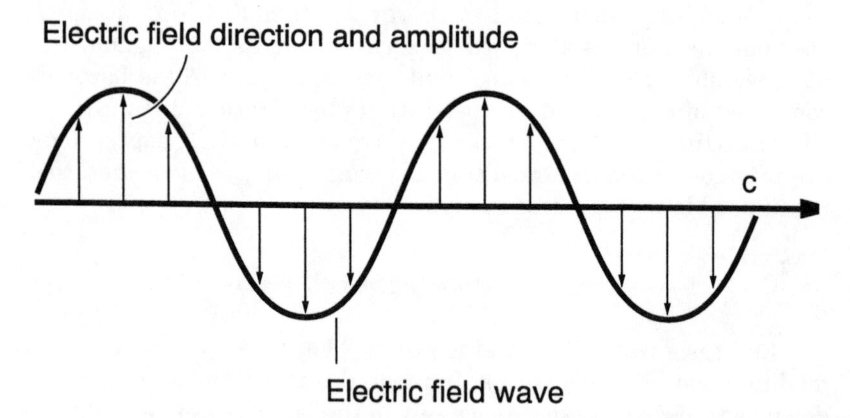

Fig. 3.1. An electromagnetic wave

1887. According to Hertz the radio waves he made with a spark gap had all the properties exhibited by light waves, which were well-known from the study of optics.

The fundamental properties of waves are *interference*, *diffraction*, and *polarization*. If we can persuade ourselves that light, and generally electromagnetic radiation, has these properties, then we shall have convinced ourselves, as did the nineteenth-century physicists, that light is in fact a wave.

Interference

If two similar waves approach each other so that their crests appear together, then the resulting wave has twice the height, or amplitude, of either wave alone. This is called *constructive interference*. On the other hand, if the crest of one wave arrives at the trough of the other, then they cancel each other, giving a zero amplitude. This is termed *destructive interference*. Both kinds are illustrated in figure 3.2.

Newton believed that light consisted of particles, which he called corpuscles. However, in 1801 Thomas Young in England performed the now-classic experiment that definitively shows that light is a wave. The experiment consisted of letting light pass between two slits that were close together, then looking for interference patterns on a screen positioned behind the slits. The slits were formed by making two parallel scratches near each other in opaque material coated on glass.

Young's experiment showed that light can give patterns of constructive and destructive interference, convincing evidence that light is a wave. This and many other experiments convinced physicists that light does not consist of particles, but of waves. A schematic of Young's experiment is shown in figure 3.3.

If we imagine throwing a stone in a pond, water waves proceed outward from where the stone hits the water. The behavior of waves follows Huygen's principle, which states that each point on a wave front may be regarded as a source for the further propagation of the wave. Huygen's principle is illustrated in figure 3.4. A new wave crest is formed by wavelets of one wavelength long whose source is a previous crest.

According to Huygen's principle, each slit acts as a source of light waves. In the case of the double slit in figure 3.3, the sources at each slit are "in phase"; that is, they are vibrating together either as crests or troughs, since they were originally parts of the same wave

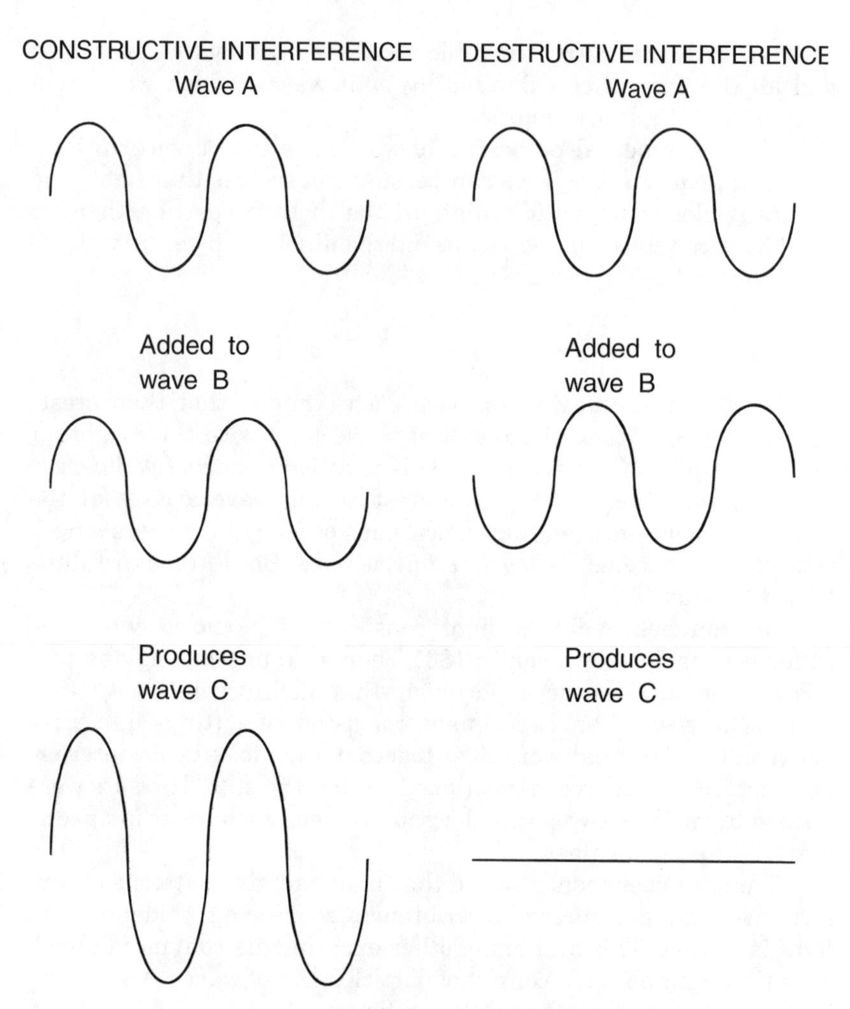

Fig. 3.2. Destructive and constructive interference

crest or trough in the incident wave. Here the incident waves (waves approaching the slits) have straight wave fronts, or crests. The distance between wave fronts is the *wavelength*, λ.

To understand how the slits act, consider the analogy of stones thrown in a pond. If I throw two stones into a pond at the same time, circular waves expand outward from each one. At points equidistant from the points where the stones strike the pond, these waves will arrive at the same time, and the resultant wave upon their arrival will be twice as high as either wave alone: constructive interference.

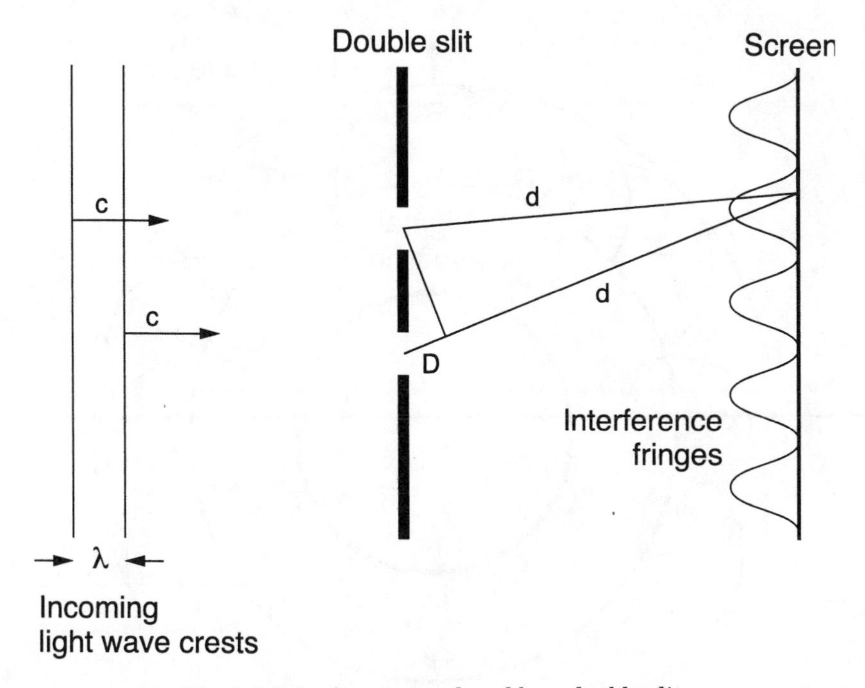

Fig. 3.3. Interference produced by a double slit

On the other hand, at another place, not equidistant, it could happen that one wave crest arrives at the other's trough and then there is no water motion at all: destructive interference.

In the case of light illuminating the double slit, the upper path length is a distance d, as shown in figure 3.3. The lower one is $d + D$. If the lower path and the upper path differ by one wavelength in going to the screen, we have constructive interference and a maximum of light intensity. Generally, if the extra distance of the lower light path, D, is an integral number of wavelengths, then a maximum of intensity will occur as shown. A special case is when D is zero. Then each path length is just d. This produces the central maximum, just as we get constructive interference at points equidistant from the two stones thrown simultaneously in a pond as just described.

On the other hand, if we imagine the light going outward such that the two paths differ by half a wavelength, or three halves of a wavelength, or generally a half-integral wavelength, we shall have destructive interference and zero light intensity. So we have alternating regions of high- and low-light intensity. The result is a series

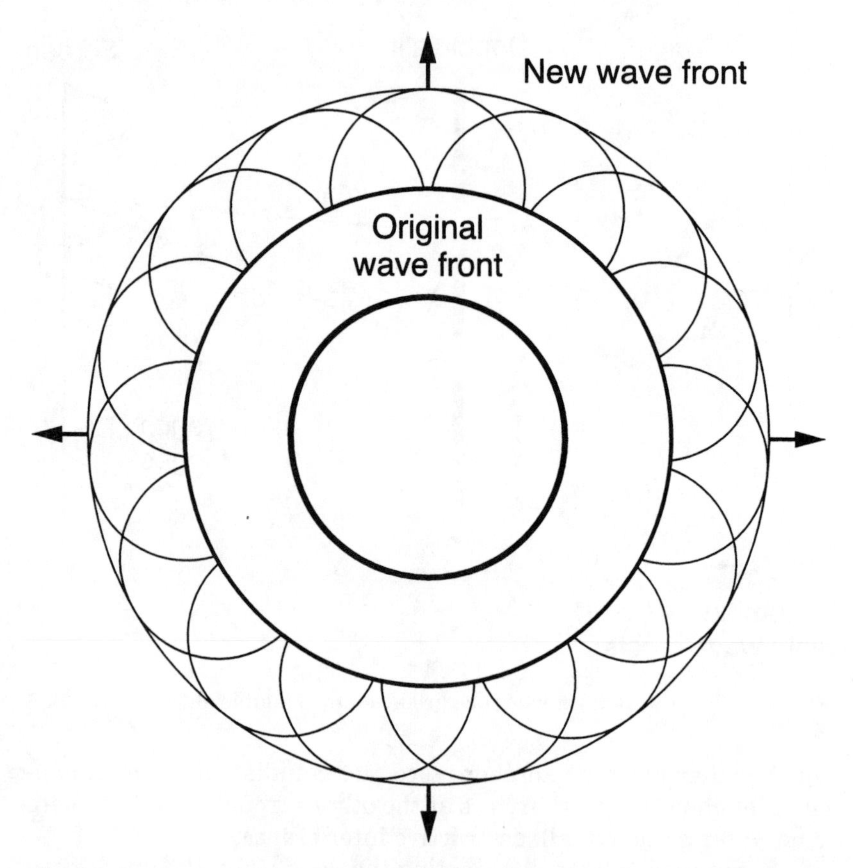

Fig. 3.4. Formation of a new wave front as described by Huygen's principle

of bright fringes. We can demonstrate the formation of such interference phenomena not only with light waves, but also with microwaves and radio waves. So electromagnetic radiation is really a wave.

Diffraction

Light waves can also interfere with themselves when they originate from the same slit rather than from a double slit. This is called *diffraction*. It provides additional evidence that light is a wave.

Diffraction can be seen most clearly when an aperture is not much larger than the wavelength of the light wave passing through it. For example, if a slit is a few times the wavelength in width, then we shall find on the screen not sharp shadows defining its aperture, but a blurred image. More familiarly, if we look at sunlight coming through a small hole, like a small hole in a sunshade, then the spot of sunlight will seem quite sharp at its edges if we view it close to the sunshade, but if we view it farther away, the edges of the spot are blurred because the sunlight has been diffracted as it goes through the hole.

Diffraction is caused by interference from a very large number of sources, for example, we can imagine that the single-slit aperture is divided up into a very large number of slits (minislits) all parallel to the aperture and all being sources of light. All of these minislit sources are "in phase," that is, they are all formed together from the incoming wave so that they will all be either in a crest or trough at the same time.

On a distant screen, light from a minislit at one edge of the aperture might form a wave crest, and light from a minislit at the middle of the aperture might form a wave trough. We now imagine that we pair off similar minislits starting from the edge and proceeding to the middle of the aperture. We have then taken the whole aperture into account. Each pair will result in a similar destructive interference on the screen. The result is that there will be no light on the screen if the path difference, D, is an integral number of *half* wavelengths.

On the other hand, if the path difference is an integral number of wavelengths, all the amplitudes of the individual minislits will add together to provide a maximum of intensity, as shown in figure 3.5. This will also occur for the central maximum where the path difference, D, is zero, just as the waves from two stones striking a pond produce a larger wave at a point equidistant from each. As we can see from the figure, the result is that the image of the slit is highly blurred. The fringe distribution on the screen is called the diffraction pattern of a single slit and was well understood in the nineteenth century on the basis of the wave theory of light.

Polarization

An electromagnetic wave is *transverse*, that is, vibrations of the wave are at a right angle to the direction of propagation. In figure 3-1 the polarization is in the plane of the paper. If a wave is plane polarized, then all vibrations occur in one plane. If the light is unpolarized, then the electric field can vibrate in any plane containing the direction of propagation.

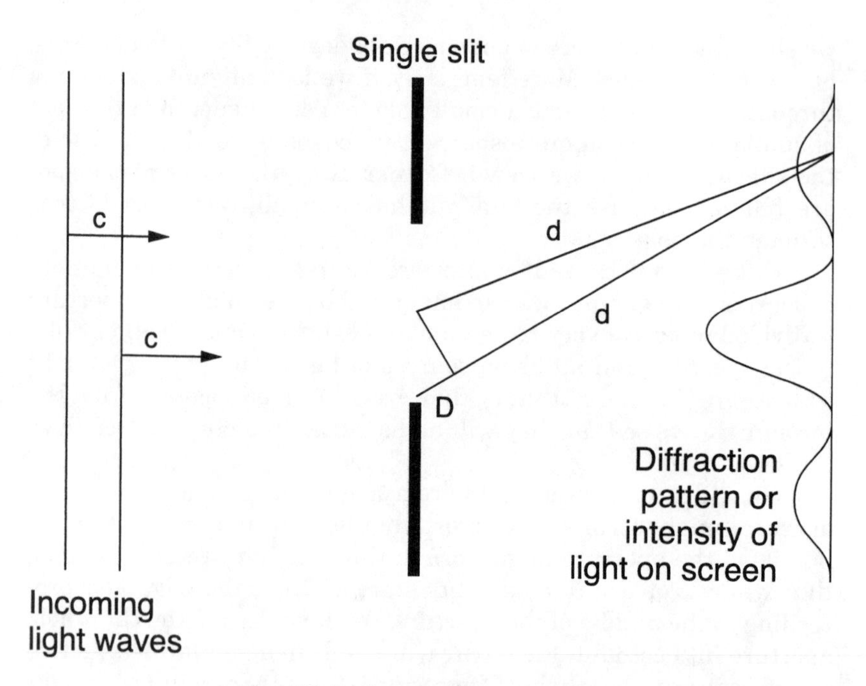

Fig. 3.5. Diffraction by a single slit

Polaroid glasses take advantage of the fact that light waves reflected from a wet roadway or snow have their electric field vibrating mostly in the plane of the horizontal reflecting surface. The glasses, via Polaroid microcrystals, accept only light whose electric field is vibrating perpendicular to that surface—that is, in a vertical direction—so the glare from the wet surface is much reduced. It is like throwing long sticks at a picket fence with spaces between the pickets larger than the width of the stick. If the stick is aligned parallel to the pickets, it will pass through, but otherwise it is blocked. Observation of polarization of light verifies Maxwell's wave theory of light and is therefore additional evidence that light is a wave.

Evidence that Light is a Particle

Given the evidence that light exhibits the properties of interference, diffraction, and polarization, light is clearly a wave. Now let's enter the twentieth century. What seemed so clear to nineteenth-century physicists now becomes a great puzzle, for sometimes light seems to be a particle! The theoretical investigation of black body radiation provided the first clear evidence of this. As we shall see in the next section, the *photoelectric effect*, the *Compton effect*, and *pair production* all are additional evidence of the particle nature of electromagnetic radiation.

Black Body Radiation

At the end of the nineteenth century, physicists agreed that electromagnetic radiation, including light, is made up of waves. There was just a small problem—which turned out to be the impetus for a great discovery. When an object is heated, it gives off first a dull red, then a bright red, and finally a white radiation. We see this in fireplaces and in incandescent light bulbs. Since Maxwell's theory of electromagnetism was so successful in many areas, it was natural to use it to explain the character of the light that would be given off by a hot object, often called a *black body*. This term emphasizes that such a body would absorb any radiation incident upon it and has no color because it reflects nothing.

A German physicist at the University of Berlin, Max Planck, had been trying for six years to predict from Maxwell's theory how much light of a given wavelength, or frequency, was radiated for a given temperature of a black body. What he found was what is now known as the *ultraviolet catastrophe*; the theory predicted that an ever-increasing, eventually infinite, amount of radiation would be emitted at frequencies beyond visible light in the ultraviolet and even shorter wavelengths. (The frequency is the number of vibrations per second of the electromagnetic wave.) This was a clear disagreement with observation.

Desperate, Planck tried a theoretical trick. He assumed that the radiation was emitted in small packets of energy, or *quanta*, which were proportional to the light's frequency. We now call such quanta *photons*. He dubbed the constant of proportionality h, and his hope was eventually, perhaps with new information, to make h go to zero. Then there wouldn't be any packets anymore and he could rid himself of this preposterous idea—for everyone agreed that light was emitted continuously, as Maxwell's theory predicted.

The assumption produced excellent agreement with experiments, but try as he might, he couldn't make h—just an ad hoc idea—go to zero. He found that *for light of frequency f, the quantum of energy permitted was a constant (h) multiplied by the frequency.*

Now called Planck's constant, $h = 4.136 \times 10^{-15}$ *eV-sec* (*eV* is an *electron-volt*).[1]

Much to his dislike, Planck was forced to keep h in his theory of black body radiation, for in fact he had discovered the idea of a light quantum. It was a fundamental discovery: the birth of quantum mechanics. It was also the first clear evidence that light also is composed of particles.

The Photoelectric Effect

In the same year that Einstein developed the special theory of relativity (1905), he also brilliantly used Planck's new idea of a light quantum, or photon, to explain the photoelectric effect. This effect occurs when light shines on a metallic surface. Under certain conditions, discussed below, it may free an *electron* from the metal. An electron is a fundamental particle that has mass and a negative electrical charge. It is probably most familiar to us as the stream of electrons that paints the picture on our television screen, or the crackle of sparks of electricity if we comb our hair when the air is dry.

In the photoelectric effect, a photon disappears into the metal and an electron is freed from it, as illustrated in figure 3.6. A part of the photon's energy is transferred *in this process* to the electron. It is an event proceeding in time. The energy and angle of emission of an individual electron is not predictable. One could say that the electron makes a decision to be emitted at a particular angle and

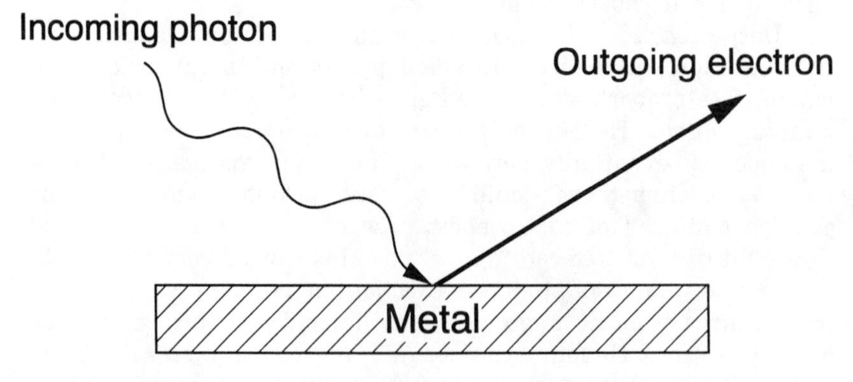

Fig. 3.6. The photoelectric effect

energy. There is an openness in that the electron makes its selection among alternatives. Thus the event takes into account its past, particularly the photon, and after its selection is made, the result is an expelled electron speeding at a chosen angle—a description very compatible with the viewpoint of process philosophy.

According to the classical wave theory of Maxwell, if the *intensity* of the light incident on the metal is doubled, then the kinetic energy (energy of motion) of the ejected electron should also double. If we imagine the light wave as a rope and the electron as a ball fastened to its end, then the more we shake the rope, the more energy of motion the electron will receive. A reasonable idea if you believe in the wave theory of light!

In fact, *when the light intensity is increased, the electron's kinetic energy doesn't change at all*. What does happen, if, for example, the intensity is doubled, is that twice as many electrons are ejected as before. Furthermore, if we shine ultraviolet light on the metal we observe electrons being released from it, but if we use red light instead, *no electrons are released at all*. These photoelectric phenomena just defy Maxwell's wave theory of electromagnetic radiation.

Einstein applied Planck's discovery of the photon to the photoelectric effect. According to Maxwell's electromagnetic theory, the energy available in a light beam is proportional to the light's intensity, but according to Planck it is proportional to the light's frequency. We therefore consider the incoming light wave to be made up of packets of energy, light quanta, or *photons*, each of value hf, where f is the frequency. Then the emitted electron would have an energy equal to hf less the energy required to release the electron from the metal.

If we use lower and lower frequencies of light, hence longer and longer wavelengths, a frequency will be reached when no electrons at all are released from the metal. The photons will simply not have enough energy to give to the electron so that it can overcome its attraction to the metal. Such would be the case for the red light discussed above.

Suppose a child is playing a game of marbles on a sandy playground. If a marble is struck, it will take a bit of energy to get it to budge once it's lodged in the sand. If the child flicks a BB shot with his or her thumb as a shooter marble, the marble will probably just sit there, unmoved. The BB just can't give the marble enough energy. Shooting more BBs will have the same effect. On the other hand, if the child uses a similar marble as a shooter, it will have much more energy of motion available to transfer to the target marble, and this marble will be ejected.

In this example, the energy required to move the marble in the sand is analogous to the energy required to remove an electron from the metal. The BBs correspond to the red photons that have insufficient energy to remove the electron, whereas a shooter marble represents a photon that has sufficient energy to eject an electron.

Classical theory would have it that if we increase the light intensity, we can give the electrons more energy so that they can escape, but this is contrary to experience. We can increase the intensity of the red light all we want, but no electrons will be freed from the metal. If, on the other hand, we consider the fact that each photon has an energy given by its frequency, then increasing the intensity of the light does not add energy to the individual photon; it only increases the number of photons. Each one still has insufficient energy to release the electron.

If the light has a frequency high enough to release electrons from the metal, then increasing the intensity will increase the number of photons; hence more electrons will be released in proportion to the number of photons incident. To give the electrons more kinetic energy it is necessary to increase the frequency of the light: use blue light or even ultraviolet light so that the individual photons have more energy to transfer to the electrons, since the energy each photon carries is Planck's constant multiplied by the light's frequency.

Einstein's explanation of the photoelectric effect as being produced by the particle aspect of light was confirmed experimentally over a range of frequencies and with different metals by Robert A. Millikan, who later became president of the California Institute of Technology. In 1921 his discovery of the explanation for the photoelectric effect earned Einstein the Nobel Prize in physics.

When physicists got somewhat used to the idea that electromagnetic radiation consisted of particles, photons, then there was a new way of considering how it might interact with matter. In particular, an American physicist, Arthur Compton, asked what would happen if a photon and an electron collided.

The Compton Effect

Arthur Compton, a physicist at the University of Chicago and another Nobel Prize winner, described the interaction between a light wave and an electron as a collision between a photon and the electron, as shown in figure 3.7.

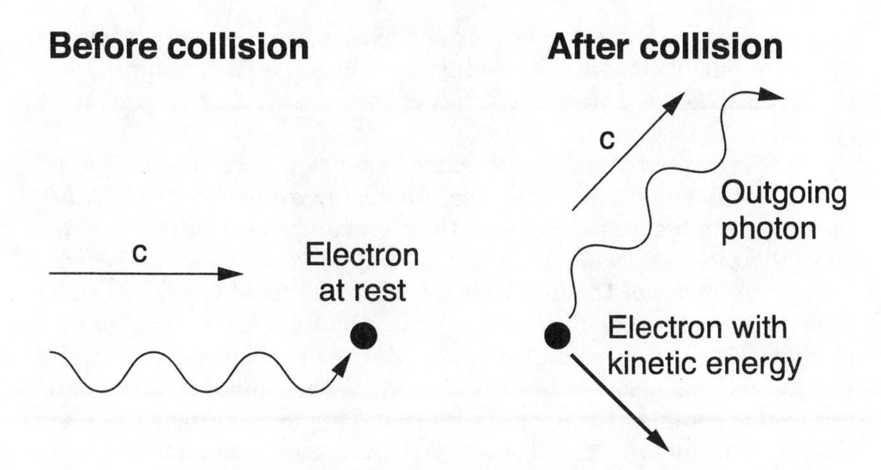

Fig. 3.7. The Compton effect

In considering the interaction of an electron with an electromagnetic wave, the Maxwell theory treats the electron as being set in motion by the electric field of the passing wave (since the electron is negatively charged and will hence feel a force from the electric field). The electron then vibrates at the same frequency as the electromagnetic wave. Again, if we consider the electron as a ball attached to a waving rope, then it will wave at the rate the rope is waving—in other words, at the same frequency as the incident wave. In so doing it creates scattered electromagnetic radiation, but at the same frequency as the incident wave.

This behavior is in accord with experience for relatively low frequency waves such as visible light. Visible light is scattered by electrons in molecules in the atmosphere without changing the color of the light. This is why the sky is blue. White light from the sun, composed of all colors, encounters air molecules. Their electrons scatter the blue light more readily than red light, and we see the scattered blue light. The opposite effect occurs when we look at the setting sun. It appears reddened, because the blue light has been scattered out of the beam of light from the Sun, and we see what remains, the red light.

However, as we approach the x-ray domain and the frequency of the electromagnetic radiation increases, the electron is removed from the atom by the incoming wave and the frequency of the scattered wave is observed to decrease—the Compton effect. This can all be readily explained by considering that the incoming x ray is a particle, a photon, which collides with the electron much like a billiard ball colliding with another at rest. By invoking conservation of energy and momentum, we can readily explain the energy acquired by the electron and the frequency shift of the photon, if we assume it is emitted at a given angle.

The scattered photon has reduced energy, a lower frequency, or a longer wavelength, and the electron acquires the energy that the incoming photon loses. None of this is explicable by the classical wave theory of Maxwell, but when the photon idea is used in formulating the theory of the Compton effect, we are able to explain that the energy lost by the photon is taken up by the recoil energy of the electron. Using conservation of momentum as well as energy conservation, we can also predict the energy of the outgoing photon in terms of the angle at which it is emitted. This process has been verified in great detail experimentally. It cannot be explained without invoking the particulate nature of light.

Here once more we have an event. Some of the energy of the incoming photon is given to the electron at rest. It is set into motion and the photon itself is transformed into a new photon—all part of an event proceeding forward in time. Again, as in the photoelectric effect, we have no way of predicting the energy or angle of emission of an individual outgoing photon. There is an openness in the process. In accord with process thought, one can think of the event making its own choice among alternatives as to what its energy and the corresponding emission angle of the new photon will be.

When photons of energy greater than one million volts became available from particle accelerators and radioactivity, a new phenomenon was observed that also requires electromagnetic radiation to be considered as a particle for its explanation: *pair production*.

Pair Production

Pair production is a process of *materialization*. It is an event that fulfills the promise of $E = mc^2$, that energy can be created from matter and also that matter can be formed from energy. Remember, energy and matter are just two aspects of the transcendent entity, mass-energy. Pair production occurs near a charged particle such as a nucleus or an electron that allows the momentum of the process to be conserved. The nucleus is otherwise unaltered in the process. In its essence a photon's energy is converted into material particles through the relation from relativity, $E = mc^2$. Since there is a "law of conservation of electrical charge" and since an electron has a negative charge, the process must create an equal and opposite charge, or positive charge, in order not to have any net electrical charge produced. Therefore, a pair of particles, the electron and its *antiparticle*, a *positron*, are created from the available energy, as is shown in figure 3.8.

The positron has the same rest mass as the electron, m_o. Thus the photon must have at least $2m_oc^2$ of energy in order to create an electron-positron pair. This corresponds to about one million electron-volts of energy (1 MeV). Such a photon is much more energetic than those of visible light, which have about two electron volts of energy. If they are created by the decay of radioactive nuclei, they are called *gamma rays*.

If we tried to explain pair production using Maxwell's wave theory, we would say that given enough visible light intensity, an electron-positron pair could be produced. But we could increase the intensity or wait as long as we liked, and there would be no pairs formed. The energy available for the materialization of pairs doesn't come from the wave amplitude, as Maxwell would have it, but from the wave's frequency, or photon energy via $E = hf$, as Planck proposed.

If the photon energy is just $2m_oc^2$, then the electron and positron will be created at rest. If the gamma ray energy is greater than one million volts, then the additional energy goes into kinetic energy of motion of the pair. If the pair has energy of motion, in a magnetic field the electron has a circular path and the positron also has a circular path, but in the opposite rotational sense. By observing such pairs produced by cosmic rays in a cloud chamber (a device that makes the particles visible) that had a magnetic field, Carl Anderson was able to discover the positron in 1932. He received the

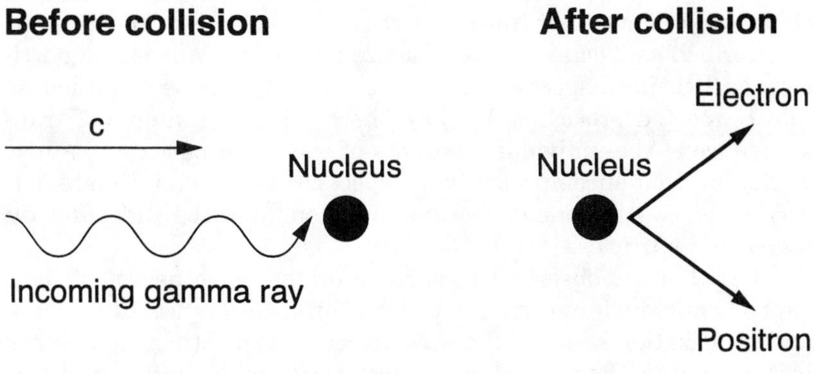

Fig. 3.8. Pair production

Nobel Prize in physics in 1936 for this research.

Today, pair production is a commonplace in nuclear and particle physics laboratories. Positrons are evanescent particles and *dematerialize* when they encounter electrons, giving rise to two photons, each of m_0c^2 or 0.5 MeV of electromagnetic radiation, called *annihilation quanta*.[2] This is again a process, the reverse of pair production. In this case material particles, the electron and the positron, disappear and are converted into the equivalent electromagnetic energy. This time $E = mc^2$ is used in its original sense of mass being converted into energy.

The interrelatedness of mass and energy and the events that are involved in transformation of one into the other are in accord with process thought where we have events proceeding in time and creating something new. Electromagnetic energy disappears and material particles, an electron and a positron, are created, or alternatively a positron and an electron dematerialize and electromagnetic energy is formed.

In the pair production process, we have direct proof of the convertibility of energy into mass. In the next chapter we shall discuss the creation of *virtual pairs* of particles in a vacuum as a consequence of the Heisenberg uncertainty relation of quantum mechanics. These pairs have only a transient existence, yet do affect our physical world. They show us that even a vacuum is filled with events—a dance of energy with particle pairs birthing and dying.

Light: A Wave or a Particle?

In view of this evidence, is light, or electromagnetic radiation, a wave or a particle? The answer is at times light exhibits a wave-like character, and at other times a particle-like behavior. Sometimes it is described as a *wavicle* to emphasize its duality. Waves and particles, each being aspects of matter or energy, can be regarded as interconnected processes. We have here another example of transcendence of the individual concepts of wave and particle to a new worldview that encompasses both aspects. We have an intimate *connection* of two phenomena originally thought to be independent: waves and particles.

Whether we observe the particle or the wave aspect of light depends on which one we are looking for. Interference, diffraction, and polarization show us the wave aspect; on the other hand, black body radiation, the photoelectric effect, the Compton effect, and pair production show us the particle aspect. The reality is always just electromagnetic radiation. We need to become accustomed to holding in our minds both the wave and particle aspects simultaneously. The microworld is simply strange to our experience that arises from the macroworld of everyday life.

Pair production forms matter from radiation, and annihilation quanta are produced by the dematerialization of matter into radiation. These phenomena are processes, events in time, and are found ceaselessly within matter as virtual processes, as we shall see in the next chapter. These concepts are in accord with process thought, in which events and becoming are fundamental.

Wave-particle duality is central to the Heisenberg interpretation of a measurement in quantum mechanics, which in turn is deeply in accord with process philosophy. As we shall see, the wave aspect can be considered as a sort of guide for the particles: They are *connected* in a fundamental way.

Matter Waves Too!

So far we have discussed the dual aspects of electromagnetic radiation: wave and particle. It turns out that this duality is not confined to electromagnetic radiation, but is a general characteristic of our physical surroundings. We are used to thinking of ordinary matter as particles. Electrons have a mass and an electric charge; energetic ones can be counted by a Geiger counter. A beam of electrons paints the picture on our TV. Yet, electrons are also waves.

Matter consists not only of small bits or particles, but also has a wave aspect. As we shall see later, the "particles" are dynamic events, not static at all. This gives us a new vision of a transcendent symmetry: Not only does light behave like both a wave and particle, but matter does as well. Since the special theory of relativity tells us that matter and energy are equivalent forms of a unity, mass-energy, if we were clever we might have expected this. In fact, in 1924 a French Ph.D. candidate at the Sorbonne in Paris, Louis deBroglie, used arguments from the special theory of relativity in formulating the revolutionary idea that matter has a wave aspect.

DeBroglie's hypothesis appeared so preposterous to his thesis committee members that they were inclined to reject it. However, deBroglie was a famous name in French nobility, with the family having served in the French diplomatic corps, so the committee looked for support for their possible rejection from the scientific community. Since deBroglie's arguments used some ideas from the special theory of relativity and since Einstein had an outstanding reputation, they sent the thesis proposal to him. His comment was probably disappointing: "It seems like an interesting idea to me." DeBroglie remains the only person who has been awarded a Nobel Prize for a Ph.D. thesis. Of course, he also received his Ph.D. degree!

DeBroglie argued that a particle will have a wavelength given by Planck's constant divided by its momentum (mass multiplied by velocity), now known as the *deBroglie relation*. We can see how the quantum concept (involving h) was taking root in fundamental ideas about the nature of matter.

For an ordinary object, such as a baseball, the wavelength is so short that we are unaware of its wave nature. For a 500-gram baseball thrown by a pitcher with a speed of 30 meters per second, deBroglie's relation gives its wavelength as 4×10^{-33} centimeters, or about a billionth of a trillionth of a trillionth of a centimeter! On the other hand, the wavelength of an electron in an atom is comparable to the size of the atom. So the electron's wave nature becomes fundamental to our understanding of this microworld.

If we use deBroglie's relation to obtain the wavelength of matter, then all of the earlier discussion of the double slit, including destructive and constructive interference and the formation of a fringe pattern, is valid for electrons as well as for light (photons), and atoms too, if we wish. We shall discuss this more in the next chapter.

In 1928 Clinton Davisson and Lester Germer at Bell Laboratories were measuring the scattering of electrons from a nickel crystal. They found strange oscillatory changes in the intensity of the scattered electrons, depending on the angle of observation with respect to the incoming electron beam. They were mystified, but when the results appeared in Europe, physicists there immediately recognized the data as experimental confirmation of deBroglie's relation. The fact that this important discovery was not first recognized in the United States attests to the vast superiority that the field of physics enjoyed in Europe at that time. Again a Nobel Prize: Davisson along with an English physicist, George Thomson, in 1937 for confirming deBroglie's hypothesis.

The deBroglie equation and the novel idea behind it created a great amount of interest in Europe and led to the development of quantum mechanics in Germany a year later in 1925. The concept of matter waves led immediately to questions such as: How do these waves propagate? How do they behave in the presence of forces? Can they be used to develop a theory of how atoms are formed? How do they interact with electromagnetic radiation? These questions, and many others, can be answered by quantum mechanics, the topic of the next chapter.

Process Thought and the Nature of Light

In the nineteenth century physicists, as we have seen, were convinced that light is a wave and that matter is a particle. In 1905, Einstein, by contrast, used the new idea of a quantum of energy to explain the photoelectric effect by assuming that light is a particle, a photon. With this new concept in mind, experimenters' goals were changed and there were subsequent further empirical proofs of the particle nature of light. This illustrates the fact that what we find often depends on what we are looking for. Goals are important; in fact, process philosophy stresses the fact that goals are essential when an event or occasion of experience forms a decision. Goals have power, not only in everyday life but in science as well.

As in the special and general theories of relativity, science of the twentieth century has shown us new *connections* between concepts originally thought to be separate: light is both a wave and a particle, and matter itself is not only a particle, but also has a wave aspect. We are invited to a transcendent conception of a wavicle for both light, or more generally electromagnetic radiation, and matter. Wave-particle duality is a fact of the physical universe. These discoveries are illustrations of the fundamental tenet of process thought that the world is formed of *interconnections*.

We have seen that when light, or more generally speaking, electromagnetic radiation, interacts with an electron in an atom (photoelectric effect) or a free electron (Compton effect), the subsequent path of an individual electron is *unpredictable*. It has an active selection among alternatives, an *openness*.

We see again that science itself is a process. Former ideas are used in an ongoing *creativity* to give us new vistas of our world. Older "laws" of physics give way to newer ones that incorporate the former ones, but deepen our understanding of the marvelous universe in which we take part. We have a *connection* between the ideas of wave and particle for light, and also a new connection that shows that matter, too, has a wave aspect.

We have seen that matter is not permanent, but under certain conditions can be transformed into energy as is predicted by the special theory of relativity. When an electron and its antiparticle, the positron, come in contact, the annihilate each other and become two photons of electromagnetic radiation, energy. This is an event, a process, wherein substances, the electron and positron, become energy.

Two process thinkers, the theologian John B. Cobb Jr. and the biologist Charles Birch, emphasize the fallacy of "substance thinking" as evidenced by physics: "Field theory, relativity physics, and quantum mechanics all point in the direction of event thinking instead of substance thinking. . . . An electromagnetic event, for example, cannot be viewed as taking place independently of the electromagnetic field as a whole. It both participates in constituting that field as the environment for all the events and also is constituted by its participation in that field."[3] That is, event thinking is holistic; the event is *connected* to previous events. Substances, by contrast, are not necessarily connected at all.

In accord with process thought, we shall find that events are the primary fact of the world. The universe is made almost entirely of electromagnetic radiation. In fact, there are about one billion bits of such radiation to every bit of matter. We are very lucky that there is one part in one billion of matter—for that is us! This radiation is often involved in events—being absorbed so that plants can grow, being emitted from glowing bodies.

Energy, not matter, is dominant in the universe, in agreement with Whitehead's view that events are primary and that simple substances are examples of "misplaced concreteness." In subsequent chapters on quantum mechanics and particle physics, we shall see further that matter itself, while appearing solid to our senses, is at the microscopic level a dynamic sea of energy exchange in what is almost entirely a vacuum—a dance of energy.

CHAPTER 4

The Microworld, Part II: Quantum Mechanics

What is quantum mechanics? Our study of relativity has given us new vistas of our familiar macroworld. Quantum mechanics is the key to understanding the microworld of atoms and nuclei, the fundamental constituents of matter. Quantum mechanics tells us how to describe the behavior of the matter waves of Louis deBroglie (discussed in chap. 3) under the influence of forces. Its applications are particularly important in describing the forces that hold molecules, atoms, and nuclei together.

Quantum mechanics is a revolutionary theory that has made possible many of the technical advances of our modern life. Since its formulation in 1925, it has never been disproved. Rather, quantum mechanics has explained the results of thousands of experiments and has led to new discoveries, such as the transistor (the heart of computers and other electronic devices) and the laser.

What has quantum mechanics to do with process thought? We shall find that it is deeply in accord with process concepts and illustrates them. The microworld that quantum mechanics describes is a realm of events and energy exchanges—of birthing and dying of particles. It is an excellent example from the physical world of the more general description afforded by the metaphysics of process philosophy, which is designed to apply to the totality of our experience.

Quantum mechanics is not usually manifest in our everyday world, since it describes the microworld. This is a world that for us is strange, even weird at times. We shall find that in the quantum world *matter is created* out of nothing and disappears immediately, making a "virtual" appearance, only to have the process repeated—a series of events. We shall find *interconnectedness* that defies the demands of relativity, and also find that individual events have an openness, a choice among alternatives that is *not predictable*.

That is not to say that quantum mechanics has no predictive quality. It does, but only for ensembles of events, not for individual ones. Furthermore, in the quantum world an observer of a physical system produces fundamental changes in it by the act of observing. The observer cannot be independent of the system being observed. All of this is very different from *classical physics*, the physics of Newton and Maxwell, which describes very effectively the physical world of our everyday experience.

These ideas from quantum mechanics are very similar to those of process philosophy discussed in chapter 1. Because in that philosophy events are seen as primary and as interconnected, interaction of an observer and a physical system would be expected. In process thought, events are unpredictable and lead to spontaneous creativity, and we shall see these features as well in the quantum world.

A Brief History of Atomic Models and Quantum Mechanics

As we have seen in the preceding chapter, the idea of a quantum of energy was first introduced by Max Planck in 1900 to explain the radiation emitted from a hot body (black body radiation). In 1905 the quantum idea was used by Einstein to explain the photoelectric effect. The next application of this concept was the brilliant intuitive work of the theoretical physicist Neils Bohr in constructing a model of the atom. We begin with a brief history of the discovery of the atomic nucleus, which led to Bohr's atomic model. Bohr's model was the historical predecessor of quantum mechanics.

Discovery of the Atomic Nucleus

By the end of the nineteenth century, it was evident from chemical studies that matter was made up of atoms. In 1897 Joseph John (J. J.) Thomson in England showed that electrons—particles of small mass and negative electrical charge—are a constituent of all atoms. This was a shock to the established belief that atoms were indivisible (the word "atom," in fact, means indivisible). Since atoms are electrically neutral, it was necessary to invoke positive electrical charges as well. This gave rise to the "*plum pudding*" *model* of the atom: a uniform sphere of positive charge in which are embedded the electron "plums."

However, new evidence was soon at hand. Ernest Rutherford, raised on a potato farm in New Zealand, had come to England to study physics and eventually became a professor at Manchester University. In the period 1909–15 at his laboratory, experiments were conducted using the newly discovered alpha rays of radioactivity, which are in reality fast-moving nuclei of helium atoms. To probe the atom, Rutherford's students, Hans Geiger and Ernst Marsden, placed an alpha-particle source in front of two defining slits so that a beam of alpha particles was directed at a thin gold foil. Alpha particles that encountered the gold atoms were detected by the flash of light they made when they arrived at a scintillation screen. This screen was coated with a phosphorescent material that gives off light when struck by a particle, just as your television screen does when a beam of electrons paints the TV picture. Figure 4.1 is a diagram of the scattering experiment.

Geiger and Marsden's goal was to find the distribution of mass in the gold atoms. Having in mind the plum pudding model, in which the positive charge is spread out over the atom, they expected the deflection of alpha particles produced in one direction by part of the atom to be largely canceled by a deflection in the opposite direction by another part. In this case the alpha particles would scatter at only small angles to the beam—like a bullet going through a swarm of mosquitoes. To their surprise they found that occasionally the alpha particles scattered at considerable angles to the incident beam, some even scattering backward.

Rutherford describes the discovery dramatically:

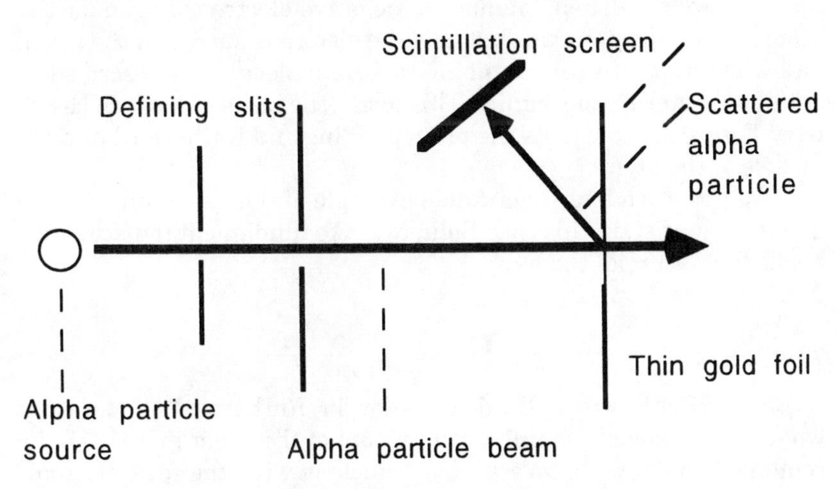

Fig. 4.1. Rutherford scattering of alpha particles

> Geiger came to me in great excitement saying "We have been able to get some of the alpha particles coming backward". . . . It was quite the most incredible event that has ever happened to me in my life . . . On consideration, I realized that this scattering backwards must be the result of a single collision, and when I made calculations I saw that it was impossible to get anything of that order of magnitude unless you took a system in which the greater part of the mass of the atom was concentrated in a minute nucleus. It was then that I had the idea of an atom with a massive center carrying a charge. I worked out mathematically what laws the scattering should obey . . . These deductions were later verified by Geiger and Marsden in a series of beautiful experiments.

This is an example of physical science at its best. An experiment is conducted to test a model. The result is unexpected, generating a new model. The new model allows a theoretical prediction, which is verified in detail by experiment. The old model is discarded. The new model, in this case of the nuclear atom, is still valid today. The supplanting of an older model by one that is in better accord with observation is a further indication of science as a process that helps us to better understand the order of the universe.

The overwhelming part of the mass of the atom, over 99.9 percent, is in the nucleus, which is positively charged. Surrounding the nucleus are a sufficient number of negative electrons to render the atom electrically neutral. Just how the electrons surround the atom, and when atoms in turn combine to form molecules, is described in detail by quantum mechanics. This forms the basis of modern chemistry. Figure 4.2 contrasts the plum pudding and Rutherford nuclear models of the atom.

We now turn to a remarkable example of the use of intuition to guide a scientist, in this case Bohr, to make fundamental discoveries about the nature of the atom.

The Bohr Atom

Neils Bohr was a Danish visitor in Rutherford's laboratory when he developed his influential model of the atom in 1913. Bohr combined the new discovery of the nucleus with the idea of quantized energy (energy in indivisible units) introduced by Planck to

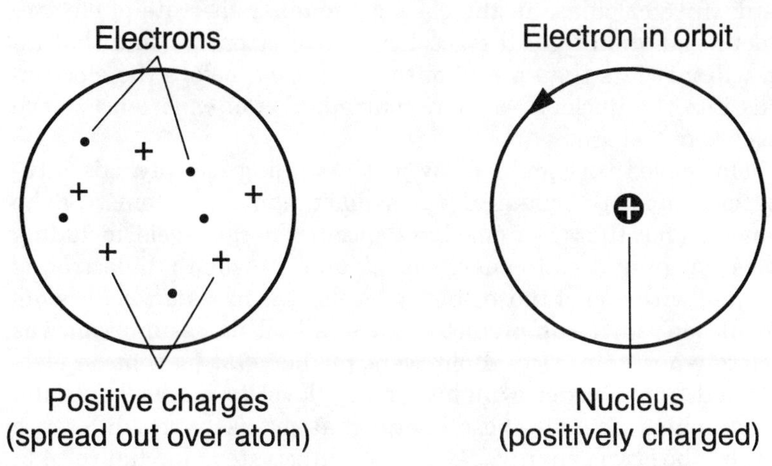

Fig. 4.2. Contrasting atomic models

produce a revised model of the hydrogen atom that agreed with experimental results to a hundredth of a percent. It also explained the periodic table of the elements. Without these spectacular results, Bohr's concept surely would have been rejected because, in a triumph of intuition, it incorporated radical ideas with no basis in classical physics.

Rutherford's experiments had shown the nucleus to be positively charged. Why don't the electrons just fall into the nucleus since they have a negative charge and would be attracted to it? Rutherford reasoned that the electron must be in orbit around the nucleus and that this motion keeps the electron and the nucleus apart in a manner similar to the orbiting of Earth around the Sun. (Earth is attracted by gravity to the Sun, but its motion about the Sun keeps it at an almost constant distance in a slightly elliptic orbit.)

For classical physicists there was an immediate problem with such a model. The problem was that according to Maxwell's well-established theory, charged particles radiate electromagnetic waves when they are accelerated. In order for an electron to remain in orbit it must be continually accelerated toward the nucleus. (If we twirl a ball attached to a string, we must supply a force on the ball by means of the string to keep it going in a circular path.) This force, according to Newton's second law, is producing an acceleration inward. If we calculate in the classical manner the rate of electromagnetic radiation from the accelerating electron, we find that an atom will last less than a millionth of a second before the electron spirals into the nucleus—a clear contradiction of experience, since atoms are in fact stable.

Bohr solved this problem by just assuming it away. His intuition told him that classical rules don't apply to atoms. So he assumed ad hoc that (1) atoms have specific energy levels, including a lowest or ground state energy level; and (2) when an electron is associated with one of these energy levels, it is in a *stationary state* and will not radiate as predicted classically. This assumption was completely new to physics. Bohr went further, stating that an electron could "jump" from a higher energy level to a lower one and thereby emit a photon whose energy was precisely the difference between the two energy levels, as illustrated in figure 4.3. Furthermore, Bohr also used Planck's quantum assumption that the energy of the released photon would be equal to its frequency multiplied by Planck's constant, h. This is sometimes called a quantum of energy.

This nonclassical electron-jumping is really an event. An electron "decides" to jump from a higher energy level to a lower one. When an individual electron will do this is unpredictable, even by quantum mechanics. It is analogous to the freedom an event has in making a selection among alternatives according to process think-

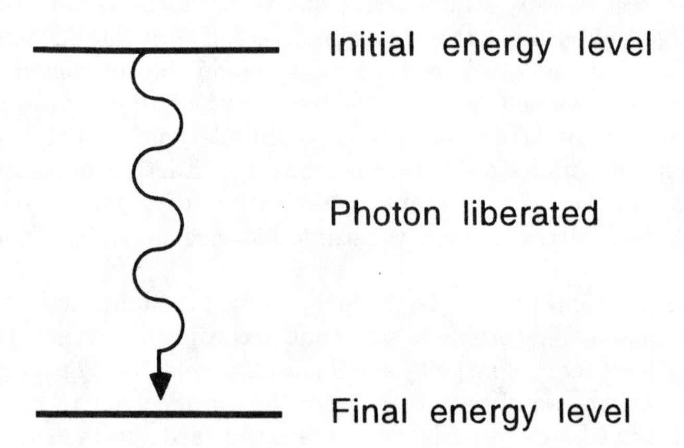

Fig. 4.3. Bohr's stationary states or energy levels

ing. The released photon is the objectified result of the selection and becomes available to influence future events—for example, to raise an electron in another atom to an excited state, or to send energy to the retinas of our eyes so we can see it.

Further restricting orbits available to the electron, Bohr assumed ad hoc that the *angular momentum*, which is the product of the electron's radius, velocity, and mass, could only be integral multiples of Planck's constant divided by 2π—that is, $h/2\pi$, $2h/2\pi$, $3h/2\pi$. . . . This was again a strict break with classical physics, in which the angular momentum could have any value. In quantum mechanics, $h/2\pi$ is used so often that it has a special symbol, $\hbar$: Using this notation, the angular momentum is *quantized* in units of $\hbar$, $2\hbar$, $3\hbar$, . . .

No one would have taken Bohr seriously except that his model explained to 0.01 percent the spectral lines observed in hydrogen. Hydrogen, as well as other atoms, does not emit electromagnetic radiation in a continuous band of frequencies, as a hot body does, but rather a discrete number of frequencies, which can be observed as lines in an optical spectograph. This had been well-known for several decades. A Swiss spectroscopist, Johann Balmer, had noticed by trial and error a curious mathematical connection among the observed lines of hydrogen in the visible region of the spectrum but had no explanation for its occurrence. Bohr's theory explained it beautifully. The spectral lines are predicted by the difference in energy between the various levels as the electron "jumps" from a higher level to a lower one.

In addition to permitting calculation of the energy levels of the hydrogen atom, the theory also predicts the radius of the hydrogen electron orbit in the lowest state, called the *Bohr radius*. It has the numerical value of $0.53 \times 10^{-8} cm$, which is in agreement with empirical evidence of atomic size from x-ray diffraction and other means. This is an important number for the estimation of the volume of an atom, which we shall refer to in chapter 5. The velocity of the electron in this orbit is about 1/137 of the velocity of light. It is noteworthy that the energy, radius, and velocity are all stated in terms of fundamental constants of physics that have been determined independently: Planck's constant, the velocity of light, and the charge on the electron and its mass.

Bohr expanded his theory to include other atoms, each of which has its own spectral signature, but it failed to predict the spectral lines correctly if there was more than one electron. It did, however, enable Bohr to predict the spectral lines of helium and lithium if they are ionized so that they have only one electron. This prediction matched some of the spectral lines emitted by the Sun, which had previously been unexplained.

Developing the idea of shells of stationary-state electrons formed by atoms in their lowest energy state, Bohr went on to apply his model brilliantly to the periodic table of the elements—for example, to explain the binding of chemical compounds and the existence of noble gases that are chemically inert.

Bohr reported these results at a "Bohr Festival" in Göttingen in 1922 with spectacular success. Einstein was moved to remark in his autobiographical notes in 1949: "That this insecure and contradictory foundation was sufficient to enable a man of Bohr's unique instinct and tact to discover the major laws of spectral lines and of the electron shells of atoms together with their significance for chemistry appeared to me like a miracle—and appears to me as a miracle today."

We have emphasized before that science is a process. Even a model as successful as Bohr's was soon replaced by one that not only included his results but also explained much more: quantum mechanics. Bohr's ad hoc assumptions, while valid, are seen to be consequences of more fundamental postulates.

DeBroglie's Matter Waves Applied to the Bohr Atom

Another important step in the development of quantum mechanics was the application of deBroglie's matter waves to the Bohr atomic model. DeBroglie's idea of matter waves that appeared in 1924, discussed in the previous chapter, had immediate application to that model. Since electrons are matter, one can ask: What are their wavelengths in the various allowed Bohr orbits? Or conversely, we can invoke the idea that the only wavelengths permitted will be those that have an integral number of waves around the circumference of a Bohr orbit. Otherwise the matter wave will destructively interfere with itself.

If there are an integral number of waves in the circumference, when a wave returns to its starting point on the orbit, it will join itself smoothly and create a standing wave. Using deBroglie's relation this procedure gives us automatically the second assumption of Bohr: that angular momenta can exist only as integral multiples of $h/2\pi$, or $\hbar$.

Formulation of Quantum Mechanics: The Schrödinger Method

The idea of deBroglie that matter has a wave aspect led very quickly to quantum mechanics. Within two years the concept of standing matter waves in atoms was seized upon by several theoretical physicists to formulate three distinct versions of quantum mechanics. Later all three were shown to be mathematically equivalent. Werner Heisenberg and Edwin Schrödinger in Germany created matrix mechanics and differential equation versions respectively, while Paul Adrien Maurice Dirac in England invented his own abstract algebra for his formulation. The Schrödinger method, generally used, is the one we shall discuss—but each has its advantages.

Schrödinger's formulation is the nearest to classical physics. It develops an equation that describes the behavior of matter waves under the influence of forces, similar in concept to Maxwell's wave equation describing electromagnetic waves. It has been criticized on the grounds that the similarity is deceptive and conceals to some extent the radical departure of quantum mechanics from classical views.

The ideas of quantum theory are indeed foreign to our usual ways of thinking. After having returned from a visit to Bohr's laboratory in Copenhagen, Schrödinger complained: "Had I known that we were not going to get rid of all this damned quantum jumping, I would never have involved myself in this business." Einstein also joined Schrödinger in rejecting quantum mechanics, even though he was a pioneer in introducing quantum concepts into physics.

Schrödinger's equation is equivalent to a statement of the conservation of energy, a concept from classical mechanics—namely that the sum of the potential energy of a particle and its kinetic energy (energy due to motion) is equal to the total energy, which is a constant. But in Schrödinger's equation the potential and kinetic energies appear as *operators* that perform mathematical operations on another mathematical function, the *wave function*. There has been a prolonged argument among the leaders in the development of quantum mechanics about the meaning of the wave function.

It was Schrödinger's position that the wave function is the amplitude of the deBroglie wave associated with the particle when it is in a field of force. That is, he asserted that an atomic wave function actually *is* the electron in its wave aspect within the atom. However, in time what came to be known as the *Copenhagen interpretation* prevailed. This concept was due originally to Max Born, a professor of physics in Göttingen, Germany, and was elaborated by

Bohr. It asserts that the wave function represents our *knowledge* of the electron, not the electron itself. Schrödinger expressed concern that this "transcendental, almost psychical interpretation" had become "universally accepted dogma."

According to the Copenhagen interpretation, the wave function, when multiplied by itself, is the *probability* of finding a particle at a certain point. *All of our knowledge of the physical system is contained in the wave function*. Thus, to ask where an individual electron *really is* in an atom, or anywhere else, is a meaningless question in quantum mechanics. The only answer available is one of probabilities.

Here a fundamental *unpredictability* enters into quantum mechanics. We speak of the probability of finding a particle in a certain place. We have no way of predicting the location of an individual particle. From the process point of view, it could be said that the state of an individual particle is not entirely determined by its past and retains some openness for selection among alternative localizations.

Concepts Resulting from Quantum Mechanics

Now let's look at a number of unfamiliar concepts resulting from quantum mechanics that are also illustrative of process thought. In the microscopic quantum world we shall find the following:

- An event at the atomic level has an openness and is *unpredictable*.
- An observer is *connected* to the system observed.
- A vacuum is filled with events—*creation* of virtual pairs.
- Particles that have interacted remain *connected*.

An Event at the Atomic Level Has an Openness and is Unpredictable

The idea that events at the atomic level are unpredictable is expressed by Heisenberg's uncertainty principle. This principle is a basic part of quantum mechanics and serves as a guide to quantum phenomena. It was introduced in 1925 by Werner Heisenberg, a German theoretical physicist, as the starting point in his version of quantum mechanics (matrix mechanics). It is particularly useful in giving us a qualitative feeling for quantum processes without mathematical complications. Heisenberg was awarded the Nobel Prize in physics for his formulation of quantum mechanics.

Generally speaking, Heisenberg's uncertainty principle states that the more we know about one complementary variable, the less we know about the other. Position and momentum (or velocity, since momentum is mass multiplied by velocity) are such a pair of complementary variables. The degree of lack of information about a quantity, such as position, is called the *uncertainty*. So if we know the position of a particle very well, we say that it has a small uncertainty, complemented by a large uncertainty in its momentum. The uncertainty principle is true generally, but manifests itself in the atomic realm where masses, hence momenta, are very small.

The scientist's way of stating Heisenberg's uncertainty principle is as follows: The product of the uncertainty in position multiplied by the uncertainty in momentum is a constant (Planck's constant divided by $2\pi = \hbar$). Since their product is a constant, the smaller one uncertainty is, the larger the other must be. So the more precisely we locate a particle, the less we know about its momentum.

To illustrate the uncertainty principle and the openness at the quantum level, we return to our double-slit experiment (see fig. 3.3). Since electrons have a wave aspect, the double slit will produce an interference pattern for electrons just as it does for light. The wavelength of the electron is precisely given by deBroglie's relation, which we discussed in chapter 3.

If we follow an individual electron as it passes through the slits, we have no way to predict where it will go on the screen. Quantum mechanics can only predict the probability that we shall find an electron at a particular place. An individual electron has a choice among alternatives, an openness. Probability is meaningless for a single event. It takes many events, perhaps thousands, to make the probability distribution evident (see fig. 4.4).

In some sense the electron makes its own choice among alternatives, subject to the constraints of overall probabilities imposed by quantum mechnics. The idea of an electron making such a selection is consistent with process philosophy that assumes the individuality and creativity of events even at the atomic level. According to process thought, each event in its history, called a *temporal society*, "prehends" previous events, including especially the previous event in that society, and then makes a selection among alternatives. These guide the destiny of the "temporal society," which at its least-developed level could be an electron.

Experiments have shown that the interference pattern is unaltered even if only one electron at a time goes through the slits. Thus the electron interferes with itself, or somehow "knows" that the second slit is open. But if we close one slit to make sure which one the electron goes through, the interference pattern is destroyed. If we consider that the electron also has a wave aspect, then this result is more readily understood, since the wave pattern from two slits is different from a single one; if we think of the electron only in particle terms, then we are mystified. We can think of the wave aspect as a "guide" for the particle aspect.

If we have a lot of electrons, we can predict the screen pattern—but only in terms of probabilities. This is one of the reasons why Einstein rejected quantum mechanics. He said in a now famous quotation: "I cannot accept the idea that God plays dice with the universe." Figure 4.4 demonstrates from experimentation how the interference pattern builds up on the screen over time with the arrival of many thousands of electrons.[1] But even as the pattern builds up, we have no way of predicting at which point an individual electron will arrive.

What we see here is an example of wave-particle duality. In order to predict the statistical pattern, we must consider the electron as a wave, yet we detect a single particle as it arrives on the screen. The wave pattern forms a sort of probability guide for the particle. Thus, the wave and particle aspects are intimately connected. They are just different ways that the electron reveals itself, depending on what observations we make, that is, we can observe the interference pattern to reveal its wave aspect, or we can detect an individual electron to show that it is a particle.

We have no way of knowing which slit the electron-particle goes through. If we try to find out by, say, shining light on it, the act of observation will give the electron so much momentum that it will destroy the interference pattern on the screen. Our lack of ability to determine the position of the electron without destroying the interference pattern is an example of Heisenberg's uncertainty principle.

In the eighteenth century, in the heyday of Newtonian mechanics and its application to astronomy, the French scientist Henri Laplace mused that he theoretically could calculate the future simply by applying Newton's equations to all the particles in the universe. The universe would be completely determined once one knew the initial positions and velocities of all the particles of matter. Quantum mechanics has changed all that; in accordance with

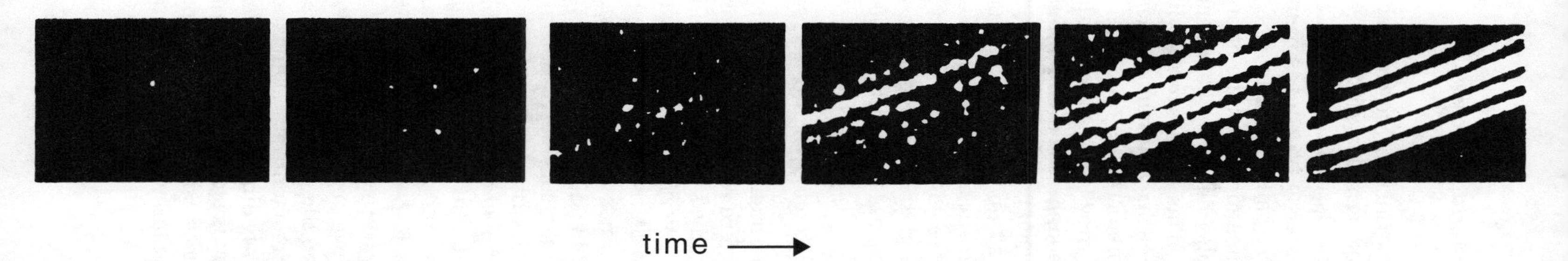

Fig. 4.4. Statistical buildup of an electron interference pattern

Heisenberg's uncertainty principle, we can never know the initial positions and velocities of particles precisely. So Laplace couldn't even get started on his calculation!

This is not just a matter of poor measurement technique; it is impossible in principle. So philosophically, the concept of a deterministic world has to be abandoned at the quantum level. We shall see in chapter 6 that it also has to be abandoned for complex systems in the macroworld. This is in accord with process thought, which holds that because each event makes a selection among alternatives, its future is not predictable.

Quantum mechanics may serve as an analogy to the idea of a dipolar God. In process thought, the divine provides overall primordial guidance for the development of the universe, but the actual development is contingent upon worldly decisions (see chap. 8). Analogously, in the double-slit experiment, the experimenter serves as a guide by setting up the slit system. However, once this experimental guideline is selected, the particle has a range of contingent choices—which fringe to select, within the overall quantum mechanical framework.

An Observer Is Connected to the System Observed

In the double-slit experiment, if we try to find the electron's vertical position, which therefore reduces the uncertainty in it, the act of observation actually changes the momentum in that direction. In the act of measuring the vertical position, we give the electron enough vertical velocity to destroy the interference pattern. Here we see that *the observer interacts with the system*. This is something new that is postulated by quantum mechanics. In Newtonian mechanics, the observer was considered objective and outside the system being measured. Rita Brock comments: "The self disappears into its objective observation of objects and pretends it has removed itself. This outmoded but still common concept of objectivity in Western thought assumes that a neutral place exists from which an observer, whose presence does not interfere with the event taking place, can tell what really occurred."[2]

Another example: If we try to measure where an electron is located in an atom, we give the electron so much momentum that it will fly out of the atom a moment later. Our knowledge of the electron's location will be greatly diminished—all we shall be able to say for sure is that the electron is located somewhere in an area as large as a football field. If we used Schrodinger's interpretation, that the electron *is* the wave function, we would have to say that after we measure its position in an atom, the electron actually physically extends over the dimensions of a football field.

If we can't, even in principle, locate an electron in an atom, we conclude that the Bohr-Rutherford orbits make no sense in quantum mechanics. Since we have no way of measuring the electron's position, the idea of the velocity of electrons in their orbits also makes no sense. Here again we have an illustration of science as process. The old idea of electron orbits, while useful in the past, is now superseded by a very different quantum mechanical picture.

So an observation of a physical system, in this case the electron in an atom, profoundly affects its future behavior. Thus, the observer and the event observed are interconnected, in agreement with the process view that events are connected to each other. (Recall that in process thought an observer is considered a complex society of events.) Quantum mechanics and process thought agree: there can be no such thing as an "objective observer" who is not involved with the events being observed. The classical idea of an experimenter who objectively observes a system without changing or interacting with it has to be discarded.

In considering the double-slit interference pattern, the wave aspect of the system determines the possibilities (e.g., the fringes on the screen), but when a measurement is made there is an "actual event" (the electron particle is detected at a particular point on the screen). This event is described by the particle nature. Out of all the prior possibilities, one is selected by the measurement. The electron's future has been altered by the measurement, since the act of measurement will greatly change its velocity. This is an intervention of the macroworld into the quantum microworld. In the act of measurement the system is altered and becomes objectified.

This is very similar to the description of a "concrescence" by an event in process philosophy, wherein the event "prehends" previous events, takes into account possibilities and goals, and then makes a selection among alternatives. At this point the event is said to be actualized and becomes an object that can be experienced by future events.

Werner Heisenberg describes the observation process:

> The observation itself changes the probability function discontinuously; it selects of all possible events the actual one that has taken place. . . . [T]he transition from the "possible"

> to the "actual" takes place during the act of observation. If we want to describe what happens in an atomic event, we have to realize that the word "happens" can only apply to the observation, not to the state of affairs between two observations. It applies to the physical not the psychical act of observation, and we may say that the transition from "possible" to "actual" takes place as soon as the interaction of the object with the measuring device, and thereby the rest of the world, has come into play; it is not connected with the act of registration of the result in the mind of the observer.[3]

Although the mind of the observer is not directly connected to the act of measurement in the Heisenberg description, it does provide for the introduction of consciousness. The assumption here is that brain function involves quantum processes. This is the position of the theoretical physicist Henry Stapp.[4] He argues that since we are able to retrieve conscious experiences, such retrieval must come from a record of some sort in the brain. The record is placed in the brain by a quantum event that is analogous to an act of measurement.

Thus brain processes involving consciousness are events. This is in accord with the event-oriented metaphysics of process thought. This description negates the duality of mind and matter that was the foundation of Descarte's philosophy.

A Vacuum is Filled with Events—Creation of Virtual Pairs

Creation at the microlevel is another process explained by Heisenberg's uncertainty principle: *spontaneous* or *virtual pair creation*. This time the pair of complementary variables are energy and time. The product of the uncertainty in energy multiplied by the uncertainty in time is again the same constant, $\hbar$.

In our everyday world we are not accustomed to a connection between two variables, for example, energy and time are independent concepts. For us it seems that if a process has a certain energy, it will not be affected by the time that the process takes. However, this is just not true. It is not readily observable in the macroworld, but in the microworld, it becomes important.

A consequence of the Heisenberg uncertainty principle is that if we are very certain about the time interval in which a process occurs, there will necessarily be quite a bit of uncertainty about the energy associated with the process. This provides energy for the process within the short time available. The shorter the time interval involved, the more energy is available during that interval. As we saw in our discussion of the special theory of relativity in chapter 2, $E = mc^2$. If there is sufficient energy available, E, then we may create a mass, m. The uncertainty principle then tells us that there will indeed be sufficient energy available, provided there's a time interval brief enough, in which case conservation of energy is violated. Energy is available from nothing!

A pair of particles is created because it is a fact of nature that electric charges are conserved, that is, an electron can't be created alone, since then a negative charge would be created. However, if a positron with an equal and opposite charge (a positive one) is created along with the electron, then no net electric charge is produced. If sufficient energy is available, nature permits the formation simultaneously of an electron of negative charge and a positive electron, or positron, of equal charge and the same mass. The particle pair is created and dies within a brief period of time in which mass-energy conservation is violated. Since the pair can only exist for a very short time, it is called a *virtual pair* in contrast to the actual pairs discussed in chapter 3. Figure 4.5 illustrates the process of virtual pair creation.

If the uncertainty in time is sufficiently minimal (about two thousand times smaller than for electron-positron pairs), then more energy is available, and a proton and an antiproton pair can be created. A proton, the nucleus of a hydrogen atom, is positively charged. The antiproton has the same mass but is negatively

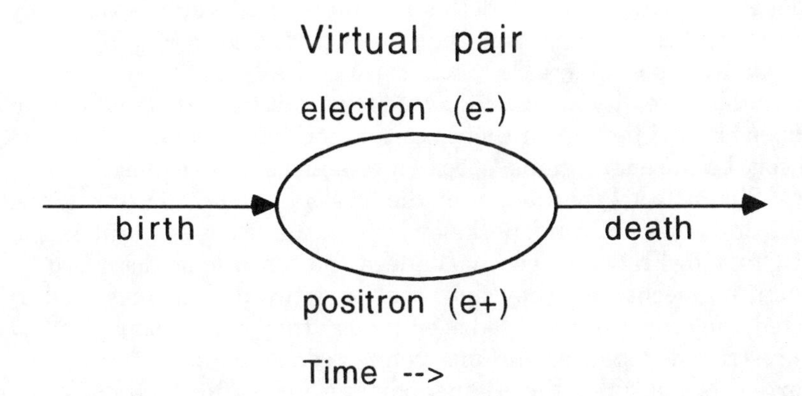

Fig. 4.5. Creation of a virtual pair in the vacuum

charged. The vacuum is filled with these pairs as well. The vacuum is also filled with neutron-antineutron pairs and so on for any pair of elementary particles.

Thus, the vacuum—the space between atoms and within atoms—is not empty, but continually and spontaneously filled with the birth and death of virtual particle pairs. This is happening in the air around us and within our own bodies. The vacuum is really filled with events—the fundamental entities of process thought.

The old idea that the vacuum is just empty space is profoundly changed by our insights from quantum mechanics arrived at through a long process of analytical reasoning and experimentation. A millennium ago the sages of China reached the same qualitative conclusion through meditation and intuition. In the world of the Tao it is said: "The great void is full of chi" or energy.[5]

Is this just a crazy idea? Why take it seriously? In physics it goes by the impressive title of *the polarization of the vacuum*. The proof is that we can calculate the energy levels of the hydrogen atom, and verify them experimentally to closer than one part in ten billion, but only if we take into account the vacuum polarization. Otherwise the calculations agree with experimentation to only one part in a million—ten thousand times less precise. So in physics it is now accepted that "this is the way the vacuum is"—a creative place indeed, and all around and within us.

Particles That Have Interacted Remain Connected

Interconnectedness is a fundamental fact in particle interactions. If two particles form a quantum state, they remain in this state even though separated by an indefinite distance. The result is that they are correlated. Einstein saw that these correlations were predicted by quantum mechanics. He concluded skeptically that "no reasonable definition of reality could be expected to permit this." In a well-known paper in 1935 he argued that quantum mechanics is an incomplete theory because it contains "spooky interactions at a distance."[6]

But in fact experiments in the 1970s by John Clauser at the University of California, Berkeley,[7] and in the 1980s by Alain Aspect at Orsay in France[8] show that indeed the world *is* as described by quantum mechanics. They show that once two particles have interacted, they remain connected even though they may be separated by unrestricted distance. The interconnectedness is such that if one particle is measured, the original partner "knows," or "prehends" the measurement and changes itself accordingly.

In Aspect's experiment two photons were detected in opposite directions from a source. The photons originated from excited calcium atoms and were correlated because they were emitted one after another in an atomic cascade. It was possible to separate the detectors to such a distance (about fifteen meters) that it was impossible for a signal traveling even at the speed of light to go from one detector to another while the interconnectedness was measured. In 1997 a group of Swiss physicists demonstrated that the quantum mechanical correlation was intact over a distance of ten kilometers using fiber optic telephone lines.[9] Thus, the correlation observed is independent of distance, at least to the extent measured, which is much beyond the constraints of the special theory of relativity.

In December of 1997 an Austrian team of physicists from the University of Innsbruck demonstrated experimentally *quantum teleportation*.[10] They were able to transport instantaneously the quantum state of a photon, in this case its polarization, from one point to another in their laboratory. This was done using two pairs of correlated photons. The initial photon whose polarization is to be transferred, which we shall call A, is entangled or correlated with one member, B, of a previously correlated pair, BC. This is done by a special measurement that determines that the polarization of photon B will be exactly the opposite of A. Since B and C are a correlated pair, that means that the polarization of C is the opposite of B. Therefore the polarization of C must be the same as A. The moment the polarization of C is measured we know instantaneously the polarization of A—the quantum state of A has been teleported. Quantum teleportation holds promise for use in computers based on quantum processes, which will in principle be faster than the present ones that rely on integrated circuits.

The empirical fact is that the world is nonlocal: that is, events in a certain space-time region can indeed affect those in another space-time region and not be constrained by the demand of the theory of relativity that no signal may exceed the velocity of light. Relativity is not violated by these interconnections since no energy or information is transferred. Nevertheless the connection or correlation exists. This interconnectedness is difficult for us to understand since there no similar experience in our macroscopic world unless telepathic contact is an analogue. Such interconnectedness is a fundamental concept of process philosophy, wherein each event is connected to all past events.

Clauser's original experiment also used photon correlations. Other experiments have been done with the bonding of two protons.

A proton has an intrinsic angular momentum, or *spin*, like a sort of elemental rotating top. When two protons get close together, their intrinsic spins align in opposite directions to each other. When this happens, they form a single quantum mechanical system. Imagine now that the two protons are separated from each other at any distance you like. If we change the spin of one proton, we shall always find that the other proton's spin will oppose it. One proton, even when widely separated, seems to "know" and to react to the spin status of the other.

This interconnection is illustrated in figure 4.6. Here two protons encounter each other with their spins in arbitrary directions. After collision and formation of the single quantum state, they are interconnected. If we now measure the spin of one proton, thereby giving it a definite orientation, we shall find the other one always

Before colliding, protons have arbitrary spins.

Spin of number 1

Spin of number 2

Velocity of number 1

Velocity of number 2

Upon collision, protons form one quantum state
with angular momenta, or spins, opposed

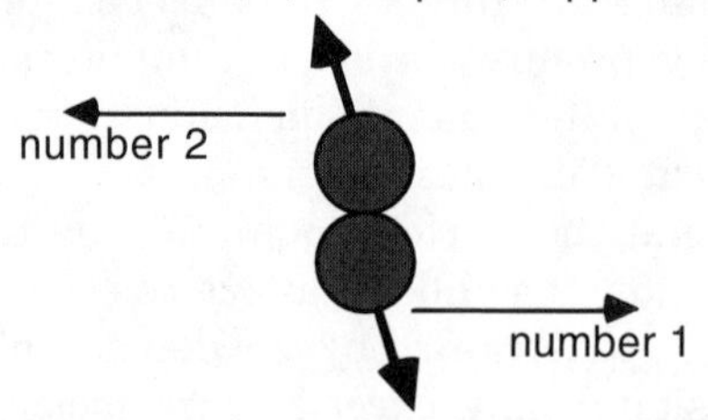

After separation, protons remain connected in one quantum state.

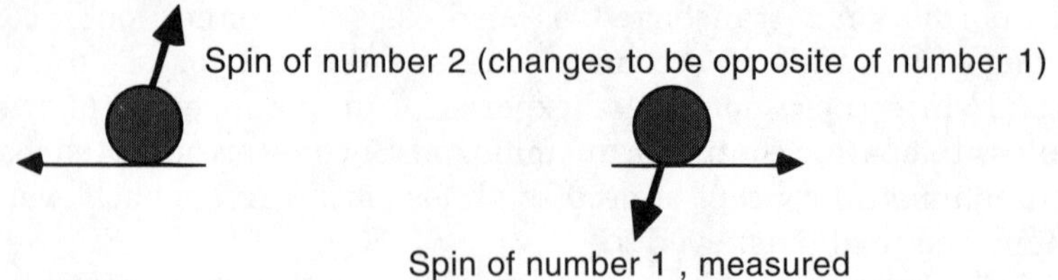

Fig. 4.6. Formation of one quantum state by a collision of two protons

oppositely directed, independent of the distance of separation—one of the "spooky interactions at a distance" to which Einstein referred.

Since quantum mechanics has never been proved incorrect and explains literally thousands of experiments, we have a choice: We can discard our belief in quantum mechanics, giving us in effect a lawless world, or we can accept its evidence that *the world is interconnected.*

We shall note the interconnection of the universe again in succeeding chapters in our discussion of particle physics, nonlinear macroscopic systems, and cosmology.

Two Applications of Quantum Mechanics

Now I shall give two examples of the quantum mechanical view of the world. The examples demonstrate how different this view is from the classical one. They show again how science is a process, with new ideas supplanting old ones. In the case of tunneling, quantum mechanics explains phenomena that are not possible in the view of classical physics. The examples show that quantum mechanics deals in probabilities: We cannot predict individual events but only the average behavior of a great number.

The Hydrogen Atom

We consider the hydrogen atom again to contrast its quantum mechanical picture with the Bohr model. The hydrogen atom is the simplest atom, and understanding it is fundamental to our knowledge of the structure of matter.

The hydrogen atom consists of just one electron and one proton held together by their electrical force. The test for any atomic theory is to compare it with the hydrogen empirical evidence. In contrast to the Bohr model, which contains ad hoc assumptions and that gives the correct atomic energy levels when only one electron is present, quantum mechanics produces the same results without those assumptions. In addition, it permits us to calculate energy levels and electron distributions when more than one electron is present, for example, it is possible to analyze the helium atom and other atoms and molecules. In fact, quantum mechanics forms the basis of theoretical chemistry and biochemistry.

In applying Schrödinger's equation to the hydrogen atom, the wave function is described by an infinite series. It turns out that *the infinite series diverges, that is, becomes infinite—unless specific values of the energy are chosen*. The energies are said to be *quantized*. Divergent mathematical behavior is not permitted because then the wave function would be largest at great distances and physically the electron would most likely be found there, which is not the case since the electron is confined in its atom. Thus the hydrogen atom's energy levels are quantized by this rather bizarre mathematical condition. The level of abstraction seems to increase as we learn more and more about the microworld.

The permitted energies are exactly the same as in the Bohr model. However, the picture we have of the hydrogen atom is profoundly different. *We really can't predict where the electron is*, only the chance of finding it at a certain point in the space surrounding the nucleus. In fact, it has a finite chance of being many Bohr radii from the nucleus, or on the other hand being much closer to the nucleus than the Bohr radius. From the process viewpoint we could say that it could decide to be anywhere within the atomic volume. This is very different from the Bohr-Rutherford model, in which the electron has a definite position in an orbit, much like a planet going around the Sun.

Surprise! According to quantum mechanics, the most likely place to find the electron is at the nucleus of the atom! The probability gradually decreases with the distance the electron is from the nucleus so there is a finite, although small, probability of finding the electron far from the nucleus. We have the image of the electron probability distribution as a fuzzy spherical ball that is most intense at the atomic nucleus and that fades away with distance. This picture of an electron probability cloud is completely different from the Bohr-Rutherford model with its electron orbits.

Figure 4.7 contrasts the quantum mechanical picture of the hydrogen atom with the Bohr-Rutherford model. In viewing this figure it is important to remember that in the Bohr theory the electron is actually confined to a two-dimensional plane, whereas in quantum mechanics, we need to imagine that the figure represents a cross section through a three-dimensional distribution of probability of finding the electron in a given atomic volume.

Suppose we ask the question: what is the probability of finding the electron at a certain distance from the nucleus? To answer this question we imagine thin spherical shells of radius r surrounding the hydrogen atom nucleus and ask: what is the probability of finding the electron in one of these shells? Since the volume of such a shell is proportional to the square of its radius, this probability is

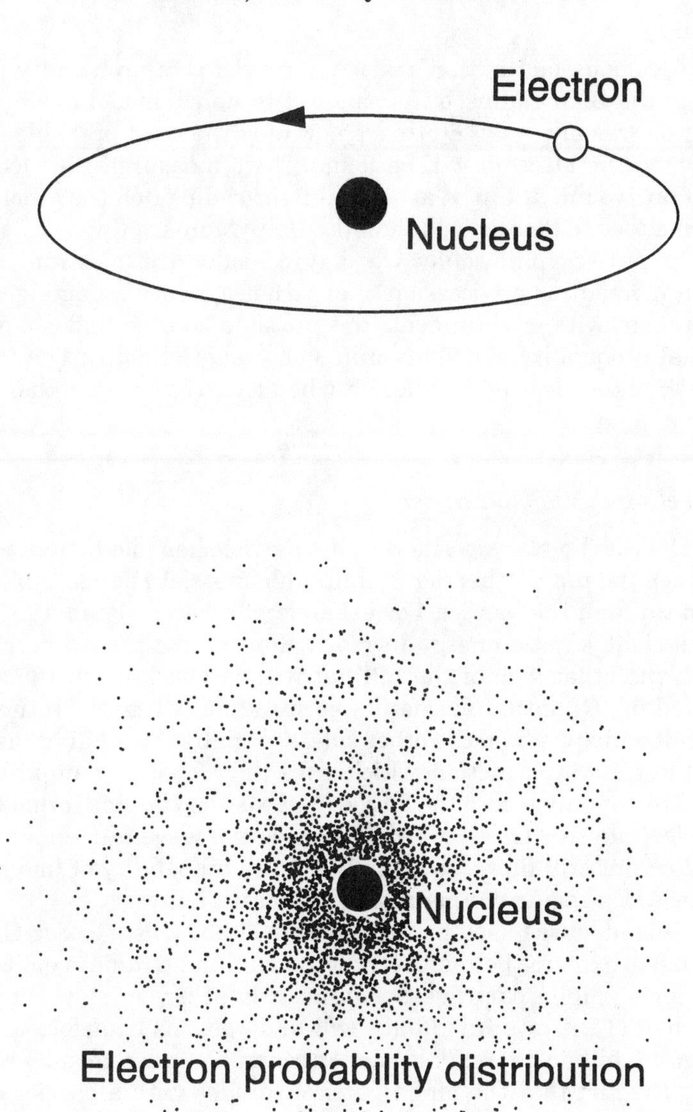

Fig. 4.7. Contrasting models of the hydrogen atom

proportional to the square of the wave function multiplied by the square of the radius.

A curious fact is that the maximum of this probability occurs just at the Bohr radius, a_o. In contrast to Bohr's model in which the electron was always at a_o, this is now only the most probable radius at which the electron will be found. If we measured the electron's position, we might find it at several Bohr radii, or on the other hand much closer to the nucleus than the Bohr radius.

In process philosophy we may consider the electron at any given moment as an occasion of experience, hence an acting entity, It prehends its environment, the protonic charge, and shapes its spatial probability distribution in a characteristic way. The characteristic of the electron is different when it is free or in another atom, or in a metal.

Tunneling through a Barrier

Classical physics posits that if a particle has insufficient energy to reach the top of a barrier of finite thickness, it is impossible for it to go through the barrier. For example, if a ball rolls up a hill with insufficient kinetic energy to reach the top, it will not be able to reach the other side of the hill but will roll back down the side it started up. Quantum mechanics permits the ball to "leak through" the hill and appear on the other side, especially if the barrier is thin. That is, the wave function, which gives the probability amplitude, is nonzero not only before the barrier, but also on the side forbidden by classical physics. On that forbidden side the wave function is much smaller, but still finite, so there is a finite probability of finding the particle in a region not permitted by classical physics.

This idea was used by the theoretical physicist George Gamow in 1928 to explain the radioactivity of uranium. Uranium emits helium nuclei, alpha particles, spontaneously at a very slow rate such that half of the uranium atoms will undergo this transformation in about 4.5 billion years. This slow rate can be understood by considering the fact that the alpha particle collides with a barrier established by the nuclear force that holds the uranium nucleus together. Although the alpha particle collides with the barrier many millions of times per second, the barrier is so impenetrable that it takes billions of years for alpha particle emission to occur. Classically there would be no penetration at all—and no explanation.

Tunneling is another example of the *unpredictability of individual events* in quantum mechanics. We have no way of predicting when an alpha particle will be emitted by a uranium nucleus. We can only say what the probability is for emission. We can calculate with some accuracy how long it will take a thousand alpha particles to be emitted, but we are ignorant of the time needed for an individual event. From the process point of view, one could say that emission occurs when the individual alpha particle has made that selection among the alternatives available to it.

Quantum-mechanical tunneling is also responsible for the tunnel diode and for the scanning tunneling microscope, which allows us to image atoms. It also makes possible the Josephson junction, our most precise device for measurement of magnetic fields. Tunneling of virtual pairs through the gravitational barrier of a black hole was invoked by Stephen Hawking[11] to demonstrate that black holes can slowly evaporate particles, leading to their eventual demise. So even a black hole is an event—albeit a very long one!

Relativistic Quantum Mechanics

Schrödinger's equation is not in accord with the special theory of relativity. This is not to say that it is not useful—it certainly is. The situation here is identical to the distinction between the classical mechanics of Newton and the new relativistic mechanics of Einstein. As long as particles with a rest mass have velocities that are small compared to the velocity of light, then Newton's equations for mechanics are correct, and Schrödinger's method for quantum mechanics is also correct. But at higher velocities, Schrödinger's formulation is superseded by more comprehensive relativistic theories that permit new insights into the nature of matter: the Dirac theory of the electron and quantum electrodyamics—another example of science as a process.

The Dirac Theory of the Electron

After the success of the Schrödinger method, theorists tried for some time to find a relativistic formulation, but there were mathematical difficulties. These difficulties were overcome by Dirac in 1928. The Dirac theory predicts what is already known from the special theory of relativity: that a free electron will have a rest mass energy, $m_o c^2$ and then in addition it can have energy due to its motion.

The new discovery in Dirac's theory was that it also predicted the creation of an electron with an equal but positive charge, a positron, with an identical rest mass, m_0c^2—the idea of an antiparticle of the electron. The positron is also permitted to have energy of motion. The theory thus predicted that if sufficient energy were available, a particle pair, an electron and positron, could be produced, each with energy of motion. Experimenters then searched for positrons in nature, and discovered them in cosmic rays in 1932. Dirac received a Nobel Prize for his formulation of quantum mechanics that included relativity.

But for all its success, the Dirac theory had limitations and also neglected an important aspect of reality: the transient existence of particle pairs that we considered earlier.

Quantum Electrodynamics

During the late 1940s, experiments with new microwave techniques were sufficiently accurate to show that the Dirac theory as applied to the hydrogen atom needed a minor correction. At this time several new formulations of relativistic quantum mechanics appeared that avoided the difficulties of the Dirac theory and at the same time gave more accurate results. One of these formulations, which is the most useful and accepted today, is that of Richard Feynman, who in his later years was a professor at the California Institute of Technology.[12]

Feynman, always the iconoclast, loved to play bongo drums and detested pretentiousness. He accepted the Nobel Prize but refused membership in the prestigious National Academy of Sciences, saying, "I am not happy being a member of a self-perpetuating honor society." He may be most remembered by the public for his critical appraisal of the *Challenger* shuttle disaster. All aboard were killed when its booster rocket exploded after launch on an unusually cold Florida morning. He generally shunned committee or administrative work but accepted this special assignment in order to help try to understand the reasons for this tragedy. To the astonishment of the blue-ribbon investigating committee, he produced a beaker of ice water, plunged into it a specimen of the rubber O-ring seal that joined together units of the rocket, and demonstrated clearly that at near-freezing temperatures the seal lost its resiliency. In his view no more talking was needed. Subsequent careful investigation proved him correct.

The Feynman formulation of relativistic quantum mechanics uses a mathematical approach that makes it conceptually easy to set up difficult problems. Working out the details may be nontrivial, however! These new formulations of relativistic quantum mechanics have come to be known as *quantum electrodynamics*, or *QED*, which is the most accurate theory that we have in physics. It permits calculation of atomic energy levels to unprecedented accuracy. QED also predicts other properties of the electron, for example, an electron is like a spinning magnetic top. QED correctly predicts the ratio of the magnetism of the "top" to its angular momentum.

The microwave experiments on hydrogen showed a shift, now called the *Lamb shift*, in the energy level of the first excited state of hydrogen of a few parts in a million compared to the prediction of the Dirac theory. The experiment was done very precisely, so that even this minor correction was determined to one part in one hundred thousand. Calculations of the shift with QED gave a result in excellent agreement. So overall the first excited state of hydrogen has been determined experimentally to one part in ten billion and confirmed by QED. This result is similar to measuring the distance from San Francisco to New York to an accuracy within the thickness of a fingernail!

Central to the calculation of atomic energy levels in QED is *the polarization of the vacuum—the creation of virtual pairs within the hydrogen atom*. The particle pairs have electrical charges that very slightly diminish the electric field between the positive hydrogen nucleus and its negative electron. So without allowing for virtual particle pairs we would not be able to predict atomic levels accurately. Such virtual pairs are at the heart of QED, and their existence is inferred from the agreement of accurate experimental results with QED's theoretical predictions. Here we have spectacular evidence that *the vacuum is filled with a dance of events*—a verification of process philosophy's concept that events rather than substances are primary, and that there is literally no empty space, in the sense of regions devoid of actual entities. What we call empty space is devoid merely of temporally ordered societies of enduring objects, such as electrons and protons.

Process Thought and Quantum Mechanics

The idea of the discrete quantum of energy, introduced by Planck to explain the radiation from a hot body, led to a whole new worldview in the twentieth century. Bohr used the quantum idea to create a highly successful model of the atom. This was followed a decade later by quantum mechanics, which is fundamental to our understanding of the microworld and to the development of technologies that have so changed our modern society.

In the nineteenth century, atoms were thought to be indivisible, but with the discovery of the electron as a constituent of all atoms and the further discovery of the atomic nucleus, a new vista opened. In the hands of Bohr the spectrum of the hydrogen atom was explained with great precision. At the heart of Bohr's theory is the idea of discrete atomic states and the further idea of events—the atomic electron jumping from one state to another. All of this is foreign to the thinking of the classical physics of the nineteenth century but compatible with process thought.

The matter waves of Louis deBroglie led immediately to a deeper understanding of Bohr's theory: electron orbits were permitted only if matter waves interfered constructively in them. In just a few months theories were developed to explain how matter waves would be influenced by forces—the birth of quantum mechanics.

Quantum mechanics has led us to a microworld in which there is an *openness*, a choice among alternatives at an elementary level, *spontaneous creativity*, and *interconnection*. These are all concepts that are fundamental to process thought.

Heisenberg's uncertainty principle shows us that there is a lack of determinism at the microlevel. This implies that the universe is not preordained but has an inherent uncertainty in its evolution. This is compatible with the process view that for each event there is available a selection among alternatives that is unpredictable.

Einstein disliked quantum mechanics precisely because its view of nature is of a statistical character with nothing to say about individual events. His statement that "God does not play dice with the universe" shows the deep religious conviction that drew Einstein away from quantum mechanics, which he regarded as an "incomplete theory."

Yet our inability to predict an individual particle's behavior at the quantum level could be interpreted to mean that a particle has a choice among alternatives, an openness, within the overall constraints of quantum theory. This behavior supports the process philosophy of Alfred North Whitehead.

We have seen that the vacuum—the space where there are no atoms or the space within an atom—is not empty. It is filled with continuous creativity—the birthing and dying of myriads of particle pairs. We have experimental evidence confirmed by a quantum mechanics that is compatible with the theory of relativity, quantum electrodynamics, that these transient particle pairs are in fact part of the world we live in. Indeed, they are part of us, within the atoms of our bodies.

This presents a dynamic concept of matter that is very different from the Cartesian view of inert matter. It might be regarded as supporting Whitehead's view that "in God's nature, permanence is primordial and flux is derivative from the World: in the World's nature, flux is primordial and permanence is derivative from God."[13] It further demonstrates the importance of events, even in so-called empty space.

Quantum mechanics is in accord with a holistic conception of the universe in that interconnectedness lies at its heart. This was another of Einstein's objections to quantum mechanics. He foresaw these "spooky interactions at a distance," which he rejected. Yet in the twentieth century there has been ample experimental evidence that particles once formed in a quantum state remain in this state and are connected to each other even though separated at great distances (the greatest at this writing being ten kilometers). This "web of interconnecting events" seen here at the level of elementary particles is also central to process thought.

Conrad H. Waddington expresses Whitehead's idea of interconnection as follows:

> In Whitehead's thought the individual character of every event, including the percipient (or perceiving event), is created by its interaction with everything else. In an act of perception, the person involved is neither a merely passive reflector nor a dominating actor who imposes his preconceived scheme of things onto his surroundings, but is instead a knot or focus in a network of to-and-fro influences.[14]

In the process view, the electron in the hydrogen atom continuously prehends the proton nucleus and acts to preserve a characteristic probability distribution of its location in space. The electron is indivisible, is an actual, acting event, or occasion of experience.

An assembly of electrons obeys the Pauli exclusion principle that each electron in an assembly, such as an atom, must be in its own quantum state. This principle permits us to explain the periodic table of the elements or the conduct of electrons in metals. This behavior cannot be deduced from the properties of a single electron. It is an *emergent* property of the whole.

We see this again in the interconnection that persists at large distances for particles that form a quantum state—it is a property of the whole quantum system. Quantum mechanics is holistic and opposed to the reductionism of Newtonian physics—that the whole is merely the sum of its mechanistic parts.

Quantum mechanics demonstrates that particles are not self contained and localized. Particles interconnect to form a whole and are in turn influenced by the whole. Reality is more like an organism or a community of events. In fact, Whitehead termed his process philosophy a philosophy of organism.

Birthing and dying of particle pairs fill the vacuum—a series of events. In the next chapter we shall see that the electrical forces that hold atoms together are thought of in the field theory of physics as arising from a series of exchanges of virtual photons between the electrons and the nucleus—again a series of events. Similarly, quarks that constitute the protons and neutrons of the atomic nucleus are held to each other by a series of exchanges of gluons—that is, by events. It is remarkable that Whitehead's metaphysics, formulated in the 1920s, is so compatible with these ideas, some of which did not emerge from physics until the 1940s and later.

CHAPTER 5

The Microworld, Part III: High-Energy Physics

What is the world ultimately made of? To the Greeks the fundamental entities were earth, fire, air, and water. Our response today is in terms of interacting particles that have dimensions from ten thousand to a billion times smaller than the atomic realm. Surprisingly, quantum mechanics describes this world of the extremely small very well. Later we shall appreciate the fact that there is a curious connection between the world of the very small and our understanding of the cosmos—how the universe is being made.

Although we speak of matter in this and in other chapters, we should remember that, according to the special theory of relativity, matter is equivalent to energy. So we could just as well speak of energy rather than matter. Also, most of the particles we shall discuss are unstable and spontaneously change to others, often with emission of some form of energy. This spontaneous changing of their properties is very much in accord with process thought. Their properties also depend on the entities to which they are related. For example, a free neutron changes spontaneously into a proton and other particles in about eleven minutes, but if it is in a nucleus of helium, it is stable. In process language, by prehending other actualities, the particular character of a society of events, in this case the neutron, is altered.

According to our best present knowledge of the constitution of matter, substances and objects, which appear so real to us, are much different when viewed at an elementary level. We shall see that matter is really composed of a myriad of events in a vacuum—a dance of energy exchanges by evanescent particles. In agreement with process thought, we shall find that the individual events demonstrate *creativity*, *interconnection*, and *unpredictability*. We shall see that the mass of an atom is contained in one quadrillionth of a trillionth of its volume, or less. So what is apparently a solid substance is really not what it seems to our senses. As the novelist Gertrude Stein would have it: "There is no there there!"

The process philosopher Charles Hartshorne emphasizes this point: "Physics undermines the view, fundamental to dualism and materialism, that the basic units of nature are inert and fully determined . . . [p]hysics now shows nature to be most fundamentally a complex of events, not enduring substances."[1] Again, the theologian B. E. Meland notes: "What has come about in the shift of imagery in the new physics and in emergent thinking generally represents not so much a reaction as a radical reconception of fundamental notions, altering the modern consciousness itself."[2]

Constituents of Matter

To appreciate the fact that substances are not at all what they seem, here we shall look at the smallest dimensions of matter that we are capable of discerning. We leave behind our familiar macroworld to investigate the bits of matter that are part of its formation. Aggregations of atoms into molecules are essential for life forms, but here we look into the nature of atoms themselves, and further into their constituents. We shall find that what appears to us as solid matter is really a dance of energy events and interconnections.

In this chapter the term *particle* will be used extensively because it is in common parlance. "Particle" has historically denoted inert bits of matter, and "elementary particle" the ultimate of such bits. Here we shall see that the particles are part of a sea of energy exchanges, and those that contain almost all the mass of the atom (quarks) are not stable. Furthermore, a particle can be viewed as a singularity in the metric of time-space, as we saw in chapter 1. Thus we are led to a view of "particle" as an entity that is quite different from that of nineteenth-century physics, even though that viewpoint persists in our culture.

Atomic Nuclei

As was described in chapter 4, experimental evidence from Ernest Rutherford's laboratory in England using a scattering of alpha particles by atoms showed dramatically that almost all of the mass of an atom is concentrated in its nucleus. Atomic dimensions are about 10^{-8}, or one hundred millionth, of a centimeter. Nuclear dimensions, however, range from ten thousand times smaller (10^{-12} centimeters for uranium) to one hundred thousand times smaller (10^{-13} centimeters for hydrogen).

Nuclei are formed from *protons* and *neutrons*. Neutrons have almost the same mass as protons but do not carry an electrical charge. A name that refers to either a proton or a neutron is *nucleon*. As we have seen, a hydrogen atom is made from a proton, which carries one unit of positive electrical charge, and an electron, which carries one unit of negative electrical charge. Thus, the hydrogen atom is neutral electrically. The attraction of the positive charge of the proton and the negative charge of the electron forms the hydrogen atom.

The nuclear electric charge, given by the number of protons in the nucleus, determines that atom's chemical nature. This is called the *atomic number*. Some nuclei have the same number of protons but differ in the number of neutrons. Such nuclei are called *isotopes*, for example, *deuterium*, sometimes called heavy hydrogen, is an isotope of hydrogen. It behaves chemically as hydrogen but has a nucleus that consists of a proton plus a neutron instead of just a proton. Deuterium is important in the nuclear reactions that furnish stellar energy. According to the Big Bang model, it was created in the initial stages of the formation of the universe. Therefore, the measured abundance of deuterium now present in the universe is an important consideration in that model, as we shall discuss in chapter 7.

The next element in the periodic table is formed when there are two protons in the nucleus. This is helium. The most prominent isotope of helium contains two neutrons as well. It is denoted as helium-4, or ^{4}He. The "4" refers to the total number of nuclear constituents: neutrons and protons. This is called the *mass number*. To form a neutral helium atom there must be two electrons to balance the positive charge of the two protons in the nucleus. If both electrons are removed (so that the atom is said to be *ionized*), then we have remaining just the helium nucleus, denoted as an alpha particle (α particle). Helium is also important cosmologically, for example, the nuclear reactions in our Sun "burn," or combine, four protons, or nuclei of hydrogen atoms, to form a helium nucleus. (This process changes two of the protons into neutrons.)

Constituents of Protons and Neutrons—The Up-and-Down Quarks

Recently we have been able to probe more deeply into the nature of matter, and we now have some answers to the question: What is inside protons and neutrons? A series of experiments at the Stanford Linear Accelerator Center (SLAC) and elsewhere in the late 1960s demonstrated that protons and neutrons are made up of pointlike objects, that is, these objects are smaller than our means of measurement can detect—at least ten thousand times smaller that the size of the proton or neutron—so we consider them points in practice. They are called by the quixotic name *quarks*. This name originated with the theoretical physicist Murray Gell-Mann of the California Institute of Technology, who postulated their existence before the experiments were performed. He is said to have named them in a moment of whimsy from "three quarks for Muster Mark" in *Finnegan's Wake* by James Joyce.

These experiments were very similar in concept to those of Rutherford described in chapter 4. This time, a beam of very high-energy electrons was used as a probe to look for the charge distribution within the proton. One can "probe" the proton to within a distance corresponding to the deBroglie wavelength of the probing electrons. That wavelength gets smaller as the electron energy increases. Therefore, as higher-energy electrons became available with advancing accelerator technology, physicists could investigate ever smaller dimensions of matter. The wavelength of the SLAC electrons was about ten times smaller than the diameter of a proton, which is about 10^{-13} centimeters, so researchers could readily probe the protonic charge distribution.

The SLAC experimental results showed that the charge in the proton is not uniformly distributed but is located in pointlike objects. The similarity to the results of Rutherford's experiments, which refuted the plum-pudding model of the atom and located the charge of the atom in the tiny atomic nucleus, is striking.

Two types of quarks are sufficient to account for the charge and mass of a proton or a neutron. They were given the arbitrary names "up" and "down," but these names have nothing to do with spatial direction. Each quark has about 1/3 of the mass of a proton or neutron. The up quark has an electric charge of $+ 2/3$ (if the charge of an electron is -1) and the down quark a charge of $-1/3$. Two "up" quarks and one "down" quark form the proton with a total charge of one. On the other hand, one up quark and two down quarks form the neutron with a total charge of zero. Since the proton and neutron are each composed of three quarks, they have very nearly the same mass.

Quarks have the unique property of *confinement*. This means that they are confined within the nucleon that they constitute. It has not been possible to observe a single quark. If we try to extract a quark by bombarding a proton or neutron with a high-energy particle, the kinetic energy of the particle is transformed into producing quarks in pairs, called *mesons*, or electromagnetic radiation. The forces that hold quarks in the nucleon or mesons get stronger and stronger as the quarks are separated making this procedure very difficult.

At the present time physicists are working with a new generation of accelerators that will produce nuclei of atoms with energies much greater than heretofore available. They hope to produce for an extremely short time a *quark-gluon plasma*, in which quarks would be separated from their nucleons and mesons and exist briefly as they presumably did in the first instant in the Big Bang before nucleons were formed (see chap. 7).

As we shall see later in this chapter, if a neutron is free, it is radioactive: It decays spontaneously to a proton and other particles. But if a neutron is in the nucleus of an atom of helium, it does not undergo this spontaneous transformation. In Alfred North Whitehead's language, the quark constituents of the neutron, the other neutron and the two protons of the helium nucleus actively prehend each other to form a society with its own defining characteristic, the helium nucleus, which is evidently stable.

In the previous chapter we saw similarly that the electron probability distribution in a hydrogen atom is special to the interaction the electron has with its proton nucleus: It prehends its environment and assumes a special defining characteristic.

The Standard Model

Particles formed from three quarks, such as protons and neutrons, are called *baryons*. In addition to neutrons and protons, other baryons have been found, and these contain new types of quarks. Four quarks besides the up-and-down quarks have been discovered and given the quaint names "charm," "strange," "bottom," and "top." The last, the top quark, was found in 1995.

Atoms in an excited state decay to their ground state by emission of a photon. In an analogous manner, protons and neutrons have been found to have excited states that are formed from baryons made from up, down, charm, and strange quarks. They decay to protons and neutrons by emission of photons as well as other particles. In this way these baryons form a web of *connections*. Their connections and their impermanence are corroborations of process thought.

According to present knowledge, all the particles of high-energy physics arise from the six quarks and from the six particles known as *leptons*. Leptons are particles that are governed by the weak interaction, described in the next section, such as electrons and neutrinos. The six quarks and the six leptons make up the three hundred or so known particles, according to what has come to be called the *standard model*.

These leptons and quarks are classified in three independent systems of "families." We have been discussing the first family: electrons, electron neutrinos, and up-and-down quarks. The particles in a family are connected in the sense that they interact with each other. For example, in radioactive beta decay, a down quark changes into an up quark, with the accompanying emission of an electron and an electron antineutrino.

The muon and its associated mu neutrino form a second family, along with the charm and strange quarks. The third family arises from the tauon, its tau neutrino, and the bottom and top quarks. The muons and tauons are similar to electrons except their masses are much larger: about 200 times and 3,500 times larger, respectively. At present we have no theories that will predict these masses, or the masses of the quarks either. All of the neutrinos described by the standard model have no mass.

We also do not understand why there are just three particle families.[3] Experiments at CERN, the European Center for Nuclear Research, show that it is unlikely that there are more than three. Table 5.1 summarizes the particle families.

Table 5.1
The Three Particle Families

Leptons	*Quarks*
(1) electron, electron neutrino	up, down
(2) muon, mu neutrino	charm, strange
(3) tauon, tau neutrino	bottom, top

The particles in these families are not further divisible to the best of our knowledge. In Whitehead's formulation of process thought they are fundamental occasions of experience.

The Discovery that Neutrinos Have Mass

For over two decades, no exceptions to the predictions of the standard model were found. Many particle physicists complained that their field had become too well understood. This changed dramatically in 1998 when it was announced that the mu neutrino must have a mass—in contradiction to the standard model. This result emerged from measurements taken in Japan in Super-Kamiokande, a 50,000-ton water tank located more than 3,000 feet below the earth's surface in a zinc mine to shield it from cosmic ray particles and equipped with 13,000 photomultiplier tubes to detect the elusive neutrino decay.[4]

Since at least one neutrino has mass, present theory predicts that neutrinos should "oscillate" from one kind to another periodically. Thus the mu neutrino may be changing to a tau neutrino during its passage through the earth. The tau neutrino would not have been detected in the Kamiokande experiment. The inferred impermanence of the mu neutrino is in accord with the process view that the universe is formed from events. If theoretical predictions are borne out, we have events happening continually as neutrinos change their characteristics spontaneously. There are several experiments now underway that should prove or disprove the idea of neutron oscillation.

Neutrinos are abundant in the cosmos, as there are about a billion for every neutron or proton. The gravitational pull of swarms of them could have effects on the distribution of galaxies. Their mass might also help to explain one of the cosmological mysteries—that there is not enough observed matter in the universe to explain its rate of expansion. We shall discuss this further in chapter 7.

The Four Forces

What holds the quarks together to form nuclei of atoms, or other baryons? What holds atoms themselves together? Modern field theory posits *force carriers*—virtual particles that are born and die by a billion trillions each second. In other words, forces are created by events. Here, at the most elementary level of matter, we find corroboration of process thought: Events are primary in the way in which the world is held together.

There are four known forces in nature: the strong force, the electromagnetic force, the weak force, and the force of gravity (in order of decreasing strength). The strong and weak forces operate only at the nuclear level, whereas the electromagnetic and gravitational forces are long range, acting in our everyday world. All of these forces result from an exchange of mediating particles that are born on one particle and die on another.

The mediating particles within the atom are virtual photons, whose creation and absorption provide the electromagnetic force that holds the electrons to the nucleus. Within the nucleus of the atom the mediating virtual particles are called *gluons*; their creation and absorption provide the strong force that holds the quarks together to form the nucleus. This is the view of modern field theory (called a *field theory* to emphasize that a field of particles is producing the force). In addition, throughout the volume of the atom, and indeed throughout space generally, a seething myriad of virtual particle pairs with very short lifetimes is, as discussed in the previous chapter, continually being formed and unformed—again an ongoing series of events.

Electromagnetic Force

The force we understand best is the electromagnetic force. According to the extremely successful theory of quantum electrodynamics (QED), discussed in chapter 4, electromagnetic forces are produced by a mediating exchange of virtual photons, for example, if we consider an electron being deflected by another electron, both negative charges, there is a repulsive electric force that changes the electron trajectory, as shown in figure 5.1. According to field theory,

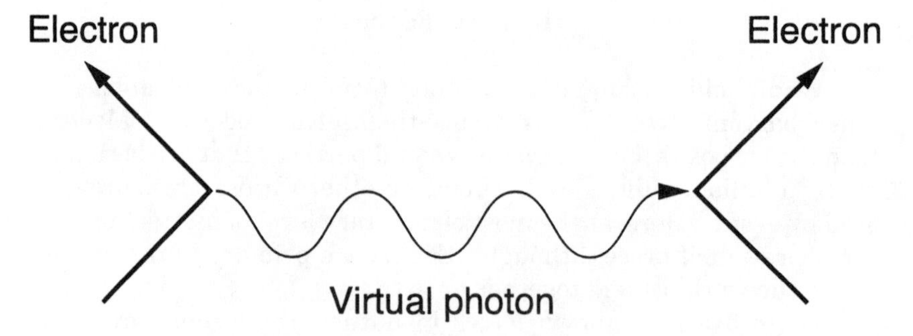

Fig. 5.1. Scattering of electrons produced by a virtual photon

this electromagnetic force arises from the exchange of a photon, which is emitted from one electron and absorbed by the other, producing a repulsive deflection in each case. If the electrical charges are positive and negative, as in a hydrogen atom, then the exchange produces an attractive force.

From the perspective of field theory, the proton and the electron that form the hydrogen atom are being held together by about a trillion billion exchanges of virtual photons per second. The most elementary atom is certainly a dynamic place! In a fantastic dance, virtual photons are born on the electron and die when they are absorbed on the proton (and vice versa) in vast numbers—an enormous amount of *creativity*. Since hydrogen is a major constituent of our own bodies, these events occur at the most basic level in each of us throughout our lives.

Strong Force (Nuclear Force)

The *strong force* is the force that holds nuclei together as well as the constituents of nuclei, the neutrons and protons. It also is the binding force of baryons and mesons. This force is mediated by an exchange between quarks of *gluons*—the "glue" binding quarks together. According to field theory, the gluon exchange produces the force that holds quarks together, just as an exchange of virtual photons creates the electromagnetic force that holds the electrons to the nucleus in an atom. The gluon exchange between three quarks, two up quarks and one down quark, which form the proton is shown schematically in figure 5.2. The quarks are shown with a size much larger than we measure. To the best of our knowledge, quark diameters are less than one ten thousandth of the diameter of the proton—sometimes called pointlike.

About a trillion trillion gluons are exchanged, born, and die, each second by any quark pair—a number of *events* impossible to imagine. Gluons are not only exchanged between quarks but also interact with each other. Within a proton or neutron there is seething *creativity* and also a web of gluon *connections* that is changing at an incredible rate. These connections form the whole, the proton or neutron, but they are particular connections that make the whole possible, as was explained in chapter 1. It is *impossible to predict* when a particular gluon will interact. It has available to it a selection for interaction among the quarks and myriad gluons within the proton or neutron: It has alternatives, an *openness*. This description of the nature of the strong force is deeply in

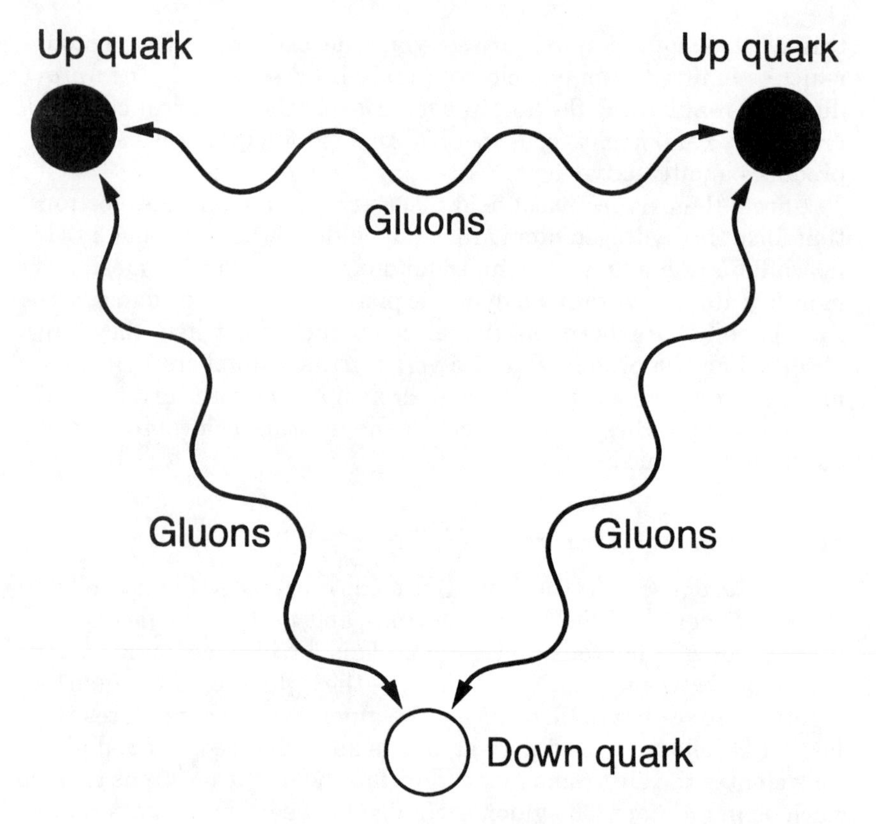

Fig. 5.2. Gluons exchanged among three quarks making a proton

accord with process thought. Quarks are in a state of nearly continuous becoming as they change their identities to and fro by their gluon connections.

Quark-quark interactions are mediated by gluons with an exchange of *color charge*. These exchanges transform a quark into another of different color. Similarly, gluons exchange color charges with each other thereby changing their identities. Color charges in this context are analogous to the electric charge that occurs in electromagnetic interactions. They have nothing to do with our ordinary experience of color. Eight different kinds of gluons are necessary to exchange the color charge.[5]

A successful theory of the strong interactions that is analogous to QED is called *quantum chromodynamics* or *QCD*. The name is given because the theory must incorporate the color charges of the gluons. It accounts quantitatively for many of the quark interactions and semiquantitatively for others.

Weak Force

A third, much weaker, basic force in nature—the one that causes radioactive beta decay—is called the *weak force*. It is responsible for the slow rate of nuclear reactions in stars so that, for example, our Sun has lasted five billion years, allowing biologic evolution to occur. According to astrophysics it will last another five billion years.

The carriers of the weak force are the $W^{\pm}$ and Z^{o} bosons. These bosons were discovered in 1984 at CERN in Geneva. This was a triumph of the *electroweak theory* developed by Sheldon Glashow, Abdus Salam, and Steven Weinberg in the 1960s, which predicted not only these force carrier particles but also their masses. They received the Nobel Prize in physics for this theory in 1979. Their work unifies Maxwell's theory of electromagnetism and the weak interactions into one theory. This unification of seemingly different phenomena is comparable in significance to the unification of electricity and magnetism by Maxwell and Faraday a century earlier. It is a further illustration of the order that we find in the universe, which is one of the fundamental tenets of process philosophy. Possible further unification of forces is discussed in more detail later.

Radioactive decay can occur by the spontaneous change of a neutron into a proton in the atomic nucleus, called beta (β) decay. Beta is a historical name, because when this phenomenon was discovered, the particles being emitted were unknown and were called beta particles. We now know that they are electrons. In addition to the electron, a particle with no electric charge and very little, if any, mass is also emitted, an antineutrino. The antineutrino is the antimatter partner of the neutrino and differs from the latter in that its intrinsic angular momentum, or spin, is along its direction of motion rather than opposed to it as is the case for the neutrino. Beta decay is caused by the weak interaction.

In radioactive beta decay, a beta particle (electron) is spontaneously emitted by nuclei. It is impelled by the weak force with an exchange of a W^{-} virtual boson. The W^{-} boson exists for only a fleeting instant during the decay, since there is not enough energy available to create it permanently. Hence, it is called a virtual boson. As an example of beta decay, free neutrons (those not bound in nuclei) decay with a half-life of about eleven minutes into a proton, an electron, and an antineutrino.

In terms of the quark model, the decay of the neutron is the spontaneous change of a down quark to an up quark by means of the emission of a W^- boson from the down quark, changing it to an up quark, and the absorption of the W^- boson by the emitted electron. This is a fundamental *creative event* of the weak interaction. In another type of beta decay a positive electron, a positron, is emitted along with a neutrino. In this case an up quark is changed to a down quark. Such beta decays occur in great numbers in producing the Sun's energy.

Beta decay is an illustration of the idea that so-called elementary particles are not permanent. Of the some three hundred known "elementary particles," only a few, such as the electron, appear to be stable. This experimental result is a striking confirmation of process thought, which maintains that an "elementary particle" is affected by its experience of other entities (with "experience," as before, considered in a broad, metaphysical sense). This combined with its intrinsic spontaneity, will limit its lifetime. As a further example, physicists have also observed inverse beta decay in which a proton and antineutrino are annihilated and a neutron and positron are created, demonstrating again the impermanence of "elementary particles." This impermanence is further evidence against the idea of changeless substances as the basic physical entities.

The neutrino is needed to carry away energy and momentum from the decay process. It is similar to a photon except that it carries only half as much intrinsic angular momentum, or spin. Neutrinos appear to have a very small mass, so that they travel nearly at the velocity of light. Since neutrinos do not have an electromagnetic field, they easily pass through matter. Neutrinos coming to us from the Sun's nuclear energy readily pass through Earth. Trillions of them pass through us every second, from below at night.

Strictly speaking, the antineutrino in beta decay is really an electron antineutrino. As we have seen, the electron and its antiparticle, the positron, and their associated antineutrinos and neutrinos are grouped under the general term *leptons*.

In both alpha and beta radioactivity, it is *impossible to predict* when a radioactive event will occur. If we have many nuclei, we can estimate the probability of when a given number will transmute, but we are ignorant of when a particular nucleus will decay. It is as if the individual nucleus decides when to do so. As suggested by the idea of occasions of experience in process thought, it evidently has an *openness* available to it.

Gravitational Force

Gravity is orders of magnitude weaker than even the weak force but has a long range and pervades the universe. It is responsible for the assembly of primordial material into galaxies and stars, and for the compressional gravitational energy that produces the high temperatures needed for nuclear reactions that fuel stars, such as our Sun. Physicists assume that the gravitational force is mediated by particles called *gravitons*, further assuming that these particles have not been observed because of their weak individual effects. We shall be discussing the influence of the gravitational force in chapter 7. Since it is so weak, it does not usually play a role in high-energy physics.

Table 5.2 summarizes the four fundamental forces.

Table 5.2
The Four Fundamental Forces

Type of Force	*Relative Strength*	*Particles Exchanged*	*Basic Theory (within its range)*
strong	1	gluons	QCD
electromagnetic	10^{-2}	photon	Maxwell's equations, QED, Electroweak theory
weak	10^{-9}	$W^{\pm}$, Z^{o} bosons	Electroweak theory
gravitational	10^{-36}	gravitons	General theory of relativity

Unification of Forces

Most physicists believe that in the first fraction of a second in the Big Bang, all the forces of nature were just one force. Gradually, as the primordial fireball cooled, the four forces as we know them were formed. We shall consider this is in more detail in chapter 7.

The electroweak theory mentioned earlier successfully unifies the electromagnetic and weak forces. As described by the theory, the unification occurs when the energies involved in the process are sufficiently high so that the rest masses of the Z^{o} and $W^{\pm}$ bosons can be neglected. At this point the bosons appear more and more like the photon (zero rest mass) that mediates the force of electromagnetism. Thus, as we conceive of very large energies, the weak force and the electromagnetic force become identical. Nature is then symmetrical in regard to these forces—either one will do. The forces have become simplified into one electroweak force.

An analogy may be useful. If we heat a permanent magnet to a high temperature, called the Curie temperature, the magnet loses its magnetism because the increased thermal motion randomizes the individual atomic domains. Pierre Curie, the husband of the more famous Nobel Prize winner Marie Curie, discovered this effect at the beginning of the twentieth century. When the domains become randomized, they become symmetrical in the sense that there is no particular spatial direction to their orientation. If we let the magnet cool, then just at the Curie temperature the magnetic domains align with each other, producing a permanent magnet again. Generally we expect that as the interaction energy increases, there will be increasing simplicity and symmetry.

A *unified theory* has been proposed that would unite the electroweak theory of leptons, photons, and wt and Z° bosons with quantum chromodynamics, QCD, which describes the quark, or strong force, interactions. The unified theory, while attractive in many respects, predicts that protons are unstable and will decay at a very small but predictable rate. Unfortunately, searches for such proton decay have been fruitless. The result is that at this time the unified theory awaits reformulation.

Undaunted, theorists now explore *grand unified theories*, GUTs. Even more speculative, these theories attempt to incorporate the gravitational force as well. According to these theories, at sufficiently high energies yet another new mediating particle would be introduced that could change gravitons to quarks, quarks to leptons, and so forth. Thus, there would be only one force at extremely high energies, and the four known fundamental forces would reassert themselves as the interaction energy diminished.

This conjectured theory is sometimes called the theory of everything. It is my view that as useful as such a theory may be, it would not explain more complex systems such as a bacterium, much less an Einstein or a Beethoven. For more information on complex systems see the next chapter.

Unified and grand unified theories are part of the Big Bang model of cosmology that we shall discuss in chapter 7. Figure 5.3 is a schematic of the four fundamental forces and these unification theories.

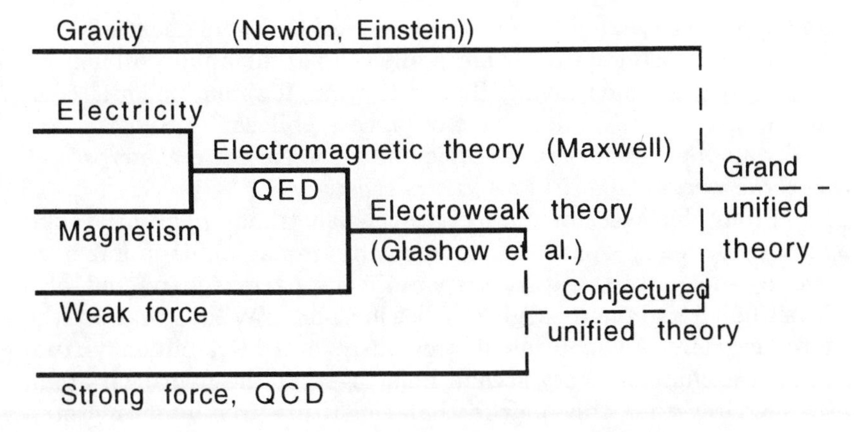

Fig. 5.3. Fundamental forces and their unification

The Emptiness of "Solid" Matter

Matter, such as a table or a chair, seems solid to us, but this is a limitation of our senses. Matter is mostly emptiness, except for a dance of virtual particles. The mass in matter occupies an extremely minuscule volume. Since we are now familiar with the constitution of matter, we are able to understand just how incredibly empty matter really is.

First, electrons have very little of the atomic mass and are points to the limit of our measurements. Second, more than 99.9 percent of the atomic mass is in its nucleus. Finally, the nucleus, as we have seen, is made up of pointlike particles, quarks. Recent experiments (at the Fermi National Accelerator Laboratory near Chicago) performed with protons that had an energy of a trillion electron volts show that the quarks approximate points to 10^{-17} centimeters or less. Since atomic diameters are about 10^{-8} centimeters, the ratio of a quark diameter to an atomic diameter is at most one in 10^{9}, or one in one billion or less.

The fraction of the atomic *volume* in which the mass of an atom is located is less than the cube of 10^{-9}. This is less than 10^{-27} of the atomic volume itself—less than one part in a quadrillion trillion, 10^{27}! A quadrillion trillion is a very large number. If we estimate all the grains of sand in all the beaches of Earth, the number is about 10^{20} (see note 23, Chapter 7); 10^{27} is ten million times larger. Matter is really empty space to an astonishing degree.

However, the "empty" atom is filled with virtual pairs and with photon and gluon force carriers that are continually being birthed and dying—a dynamic, *creative process* at the most elementary level of matter. So solid matter is not really solid at all, almost all nonsolid; and at the same time is filled with dynamic radiation, or a "complex of events" in the language of process philosophy. Substance is really an effect of our macroscopic human senses; *at the microscopic level the world is a web of a series of events*.

From this we can appreciate the fact that apparently solid matter is really space filled with an unimaginable number of events—not only those from spontaneous particle pairs but also from the virtual photons, gluons, bosons, and gravitons that are the force carriers. As mentioned previously, a trillion billion virtual photon exchanges every second hold together the hydrogen atom. To hold together the quarks that constitute that atom's proton requires the exchange, the birthing and dying, of a trillion trillion gluons every second. Thus, the world at its most basic level is really the realm of events, not substances. Hence, the fallacy of "substance thinking."

Substance thinking sees connections as external to substances. Connections are secondary—an afterthought. They are like the collision of two billiard balls. The connection in the moment of collision is momentary and unimportant. What is important are the substances—the billiard balls. As we have seen, recent physics supports process philosophy's rejection of this way of thinking.

Process Thought and High-Energy Physics

The present status of high-energy physics leaves several unanswered questions. Why are there only three particle families? Why doesn't the proton decay? How do we account for the mass of the mu neutrino that should be zero according to the standard model? Nevertheless we do have a consistent picture of the microworld when there is increasing interaction energy.

We find that what appears to be solid matter is really a vast number of mediating virtual particles being exchanged. These basic units of nature are far from inert but are ever changing and evolving in a dynamic quantum sea—a sea that is *creative*, *connected*, and with a certain *openness* at the level of individual events.

In accord with the process view, events constitute an atom to a remarkable degree. These are events made up of virtual photons that hold its electrons to its nucleus, events from spontaneous creation of pairs within it, and finally more events from gluon exchanges within its nucleus that hold it and its nucleon constituents together. All this is taking place in a void to an extremely high degree of approximation. Experimentally pointlike particles, quarks and electrons that contain the mass of the atom, occupy only one part in a trillion quadrillion of the atomic volume.

Quarks and the other entities in the standard model seem to be the world's smallest and most basic temporal societies of events. As far as we know, they are not divisible into spatially more ultimate entities. Quarks "prehend" through their associated gluons the other quarks that make up a proton or neutron. They are acting entities that are fully actual and are continually in the process of becoming by their gluon exchanges.

Similarly the electrons in atoms are held to their nuclei by exchange of virtual photons. They "prehend" their environment by a continual process of becoming through those exchanges. Since, as far as we know, they are indivisible, they also qualify as fundamental entities. Indeed, they are part of the standard model.

It should be emphasized that according to process thought, the electrons and quarks depend on their active participation and on their relationships in forming their respective atoms and nucleons. By their mutual prehensive acting, the defining characteristics of the atom or nucleon emerges. This formulation is quite different from that based on substances. In that traditional perspective, the character of the whole is just the sum of its parts. The parts are not changed by participation in compound wholes.

In process thought, interconnections among events form new events—*a web of connections*—as evidenced by the myriad of virtual photons and gluons that are birthing and dying to form the basic bits of mass-energy. In accord with the process view, particular quarks constitute a whole, such as a neutron or proton. Within the whole there are particular connections among its gluons and quarks that make that special whole possible.

Again the spontaneous changing of one particle into another is further evidence for the validity of process thought. Matter is not inert, but in process. Nothing is inherently stable and unchanging, for example, we normally think of electrons as stable. Streams of electrons paint the images on our television screens. Yet if an electron comes near a positron, it is annihilated into electromagnetic energy and ceases to exist. We could say that electrons have the *potentiality* to change into electromagnetic radiation under the proper circumstances.

Quarks are in continuous transformation through gluon exchanges; electrons and neutrinos change through interactions with other particles. In 1998 it was discovered that the mu neutrino changes spontaneously while passing through Earth, presumably into a tau neutrino. These transformations are in accord with process thought's contention that the world is composed of events in the process of becoming.

The *connections* produced by photons and gluons lead to the forces that hold atoms and their nuclei together, so that we and the world about us can exist.

Thus, the picture that emerges from high-energy physics is in conceptual agreement with the basic postulates of process philosophy, in particular that events rather than substances constitute the world.

CHAPTER 6

The Macroworld, Part I: Complex Systems

In this chapter we shall discuss a relatively new field of investigation that is changing our views of the physical world: the study of *complex systems*. Such systems are difficult to define but are characterized by the involvement of many trillions of atoms and their interactions with new laws of behavior that we are only beginning to understand. They are phenomena on a human scale. Biologic entities are complicated examples of complex systems. We shall look at a simpler one—a thin layer of heated oil. In accord with process thought, complex systems are interlinked in a web of *connections* and are *unpredictable*. They produce remarkable *self-organization* and *creativity*.

The self-organization of complex systems discussed in this chapter is an example of the creation of order and increasing complexity that is an essential feature of process thought. In process language, the subjective aim of a society of occasions of experience, or a society of events, is to maximize its satisfaction or enjoyment. Such satisfaction is more deeply felt for a complex system than for a simple one—a human being is capable of much more enjoyment from creativity than an amoeba.

According to Conrad H. Waddington:

> Whitehead insisted that an event is not merely an assemblage of numerous relations between many different things thrown together in a disorderly heap. On the contrary the various "feelings" of one thing for another are *organized* into something with a specific and individual character. . . . Organization occurs when the relations are of such a kind that they tend to stabilize the general pattern against influ-

> ences which might disturb it. That is to say, organization confers on the entity an enduring individuality which a mere assemblage lacks.[1]

This is the role of complex systems.

Entropy and the Arrow of Time

A useful concept in discussing complex systems is *entropy*. One definition of entropy is the amount of energy contained in a system divided by its temperature. Entropy increase is also the sense of the direction of time.

As an illustration, we know that if we mix a glass of cold water and one of hot water, we get two glasses of lukewarm water. But this process can't be reversed: we can't mix two glasses of warm water and get one hot and the other cold. We find that entropy increases when the cold and hot water are mixed. Time flows in the direction of mixing hot and cold water to get lukewarm water—not the reverse.

More precisely, if we consider a system that is closed so that energy can't flow in or out, then entropy increases with time, for example, we might imagine a gasoline engine running inside a well-insulated container. Some heat will be converted into useful energy by the engine, but the entropy of the system will steadily increase with time.

It has been common among modern physicists to put this point conversely by saying that the direction of time is actually constituted by the increase in entropy. For process thought, by contrast, time does not depend on this contingent process of the macroworld but is a fundamental feature of reality, being formed by the fact that all actual occasions prehend previous events. Thus irreversible time exists for individual atoms, protons, or even quarks. This point is discussed further in the last section of the next chapter.

Here is an example: Although entropy does not constitute the very nature of time, entropy is important to our lives. Without the low-entropy source of energy that is our Sun, we and all life on Earth could not exist. It is not only that we get our energy from the Sun, but also that we get a high-quality energy. The photons in sunlight have enough energy individually to produce photosynthesis. If the Sun produced the same amount of energy for us, but it was infrared radiation, heat radiation, instead, life could not exist on Earth.

Note in the example that the energy arriving on Earth from the Sun must be the same as the energy leaving Earth to space. Otherwise Earth would heat up or cool down. However, since the Sun's energy comes from a much hotter source than the energy leaving the much cooler Earth, the entropy of the energy coming to us from the Sun is much less than the entropy leaving Earth for space.

The Sun as a Source of Low Entropy

Our Sun is a prodigious source of energy. Every second it transforms 600 million tons of hydrogen into 596 million tons of helium. Four million tons of light energy (from $E = mc^2$) are radiated out into space every second. Earth receives only a small part of this energy, but it is essential to life. Equally important, this energy is of high quality. It is emitted in visible, low-entropy light that plants can use.

When we discuss the Big Bang in the next chapter, we shall see that it somehow created a low-entropy universe that permits life to exist. Why is the universe formed this way?

All plants and animals, including ourselves, are highly ordered, complex systems of low entropy. They are open systems and do not violate the second law of thermodynamics. If we consider all the heat output and excreta of our bodies to the environment—all highly disordered—then overall (viewed as a closed system) the entropy associated with the existence of a human being is increasing, but viewed by itself, a human is a low-entropy, highly ordered system. Where do we get the low entropy? Directly, by eating low-entropy plant or animal foods, which, before they are eaten, are highly ordered.

Ultimately, our source of low entropy is the Sun, which is extremely hot and therefore a low-entropy source for a given amount of energy. This energy arrives on Earth in the form of visible light and through the miracle of photosynthesis is transformed into highly ordered, low-entropy plant material—food for us either directly or through the animal food chain. Figure 6.1 illustrates this and calculates the entropy change for a given amount of energy that is received from the Sun. Since Earth is not getting colder or hotter on the average, it is in equilibrium: The same amount of energy must leave Earth by radiation to space as is received from the Sun, Q.

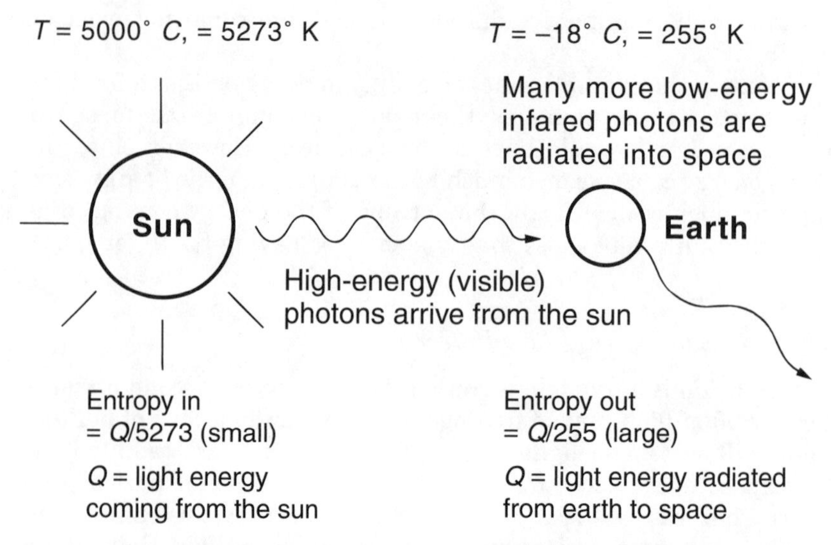

Fig. 6.1. The sun as a source of low entropy for Earth

In the figure, entropy is calculated from the *absolute* temperature, *K*, which is the temperature in degrees centigrade plus 273 degrees. (At absolute zero all molecular motion ceases.) Short wavelength photons of high energy arrive from the Sun, and many more longer (infrared) photons are radiated from Earth to space.

The temperature of Earth as viewed from space is −18 degrees centigrade. Earth's surface temperature is on average +15 degrees centigrade due to the greenhouse effect of the atmosphere. Note that the entropy of the energy leaving Earth is much larger than the entropy of the same amount of energy, *Q*, arriving from the Sun—in accord with the second law of thermodynamics (see the section after next).

Boltzmann's Alternate Definition of Entropy

In trying to understand the "arrow of time," the nineteenth-century Austrian theoretical physicist Ludwig Boltzmann discovered a new way to discuss entropy. His alternate description states that the amount of entropy is proportional to the number of ways in which a system can form itself, a measure of the degrees of freedom of a system—so that highly ordered systems have less entropy. *So entropy is a measure of disorder in a system*, for example, in a cup of water the molecules are free to move about in random paths, but if the cup of water is frozen, the water molecules have only definite positions in a crystal lattice. So a given amount of ice compared to water is a more ordered system and has less entropy.

If we consider all the waste products that a life form produces in order to survive, then we have to say that the life form increases the overall entropy of the world. However, if we focus on the living organism itself, we shall find that it produces more order and complexity and hence has less entropy than its constituents from the environment. The anthropologist Gregory Bateson liked to describe the struggle of life forms to build complexity and order—producing negative entropy—as the *sacramental*. This is the creative life force.

The Second Law of Thermodynamics

The *second law of thermodynamics* can be stated in this form: In a physical process the *entropy in a closed system remains the same or increases*. With a given amount of energy in a closed system the second law mandates that the temperature either remains constant or decreases. This implies the "heat death of the universe" when at last everything comes to a final equilibrium at low temperature.

In a closed system, if the entropy change in a process is zero, it is said to be a *reversible process*. Only idealized systems are reversible. In the real world of complex systems, there is no such thing as a reversible process, so entropy always increases. A real engine or a living organism has an *irreversible process* and is a *dissipative system*. Energy is dissipated or lost for useful work in order to maintain the system. Here we must consider not just the individual or engine, but also the environment that furnishes the heat energy to drive the system.

The second law of thermodynamics, although very powerful, is valid only for systems very close to equilibrium. In equilibrium there is no flow of heat at all; the system is static. If systems are far from equilibrium, new laws apply. During the last two decades discoveries by the Nobel Prize chemist Ilya Prigogine and by others have helped us to begin to understand nonequilibrium, or dissipative, systems. They have the ability to self-organize into highly complex and ordered structures. As entities they have less entropy than their disassembled components: Their creation opposes the second law of thermodynamics if we neglect their effect on the surroundings.

The Arrow of Time

The equations of quantum mechanics, Newton's equations, and Maxwell's equations are all time reversible; that is, these "laws" of physics don't change if the time, or t, is replaced by $-t$, as in a movie running backward. However, this does not mean—contrary to the hasty conclusion often drawn—that the physical realities that these equations describe are themselves time reversible. In any case, for *calculations* that use these time-reversible equations, it is necessary *in practice* to use boundary conditions or initial conditions that produce an intrinsic time asymmetry. Boundary conditions select the region in which the calculation is made. Initial conditions set the values of the other variables in the calculation when time begins.

In our experience, time flows asymmetrically from the past through the present to the future. In the biologic world, individuals, including ourselves, are born, grow older, and die. This gives us the intuitive sense of the arrow of time. In our discussion of cosmology, we shall find that the universe itself is evolving as stars and galaxies are born and die, and that the composition of the universe itself changes with time.

Another example of time asymmetry, attributed to Boltzmann, is to consider the ripple produced by throwing a rock into a still pond. The wave expands outward and reaches the edges of the pond. Now imagine trying to time-reverse this situation. It would be necessary to coordinate precisely all the almost infinite disturbances at the pond's edge so that they would move inward, grow in amplitude, and converge on a single dimple. In practice, this is an impossibility.

For processes in the macroworld that involve complexity, then, such as a large group of interacting molecules, there is clearly a direction of time.

Some physicists who have a reductionistic viewpoint assume that the microworld, in which entropy does not obtain, has no arrow of time. They assume that it is completely described by the time-reversible equations of quantum mechanics and electromagnetic theory. For these physicists, the whole is the sum of its parts. For them, there should be a "theory of everything" that starts from the fundamental particles and interactions discussed in the previous chapter and that enables us to understand the entire universe. But if the theoretical constructs in the microworld are time reversible and if those in the macroworld are not, this creates a basic difficulty in building a macroworld from the microworld.

As we shall discuss further in the next section, Prigogine and others argue that reversibility is a theoretical construct and that *real* systems are always irreversible—hence we always have an arrow of time. Similarly, Roger Penrose, a theoretical physicist, argues that a correct theory of quantum gravity (a theory not known at present but that would describe both quantum mechanics and gravity) should be time irreversible.[2]

This point of view has received additional corroboration recently from experiments that have shown that even at the most fundamental level there is a distinction between going forward and going backward in time. Two similar experiments were performed at Fermilab, in Batavia, Illinois, and at the European Center for Nuclear Research (CERN) near Geneva, Switzerland.[3] At CERN, colliding protons and antiprotons formed *K* and anti-*K* mesons, which can transform into one another. The scientists found a slight, but significant, difference in the relative numbers of these mesons, meaning that if time were reversed, the relative numbers would reverse. In the Fermilab experiment, a beam of *K* mesons gave a pattern of particle tracks whose shape would vary if time were reversed. Thus, there is a distinction between going forward and going backward in time, meaning that the latter is not really possible.

Time irreversibility is fundamental to process thought, wherein one occasion of experience succeeds another. The past is composed of events that have occurred and that contribute to the occasion of experience that is now occurring. The occasion makes a decision in accord with its subjective aim and becomes an object of information to be prehended by future events. The future has as yet no occasions.

David Ray Griffin, the process philosopher, argues that in the process view, time is irreversible even for the most elementary entities: "But if we can talk about 'physical experience,' about the presence of at least some iota of experience at even the most elementary level of nature, then we can say that time is real there too."

> The notion that each event prehends previous events—and this feature, that prehension is always of <u>antecedent</u> events, is fundamental—gives us not only time's asymmetry, but also its irreversibility. A prehension should not be thought to be simply a primitive form of sensory perception, at least if sensory perception is thought to involve merely a <u>representation</u> of an external thing. Rather, prehension involves an actual grasping of the prehended object, so that the object is included within the prehending experience. This means that, insofar as we speak of the prior, prehended event as the cause and the prehending experience as the

effect, the cause has literally (if only partially) entered into the effect. This gives time its irreversibility in principle.[4]

Nonlinear Systems

Complex systems are nonlinear. Nonlinearity can produce surprises in predictions, may make them impossible, and can lead to chaos, for example, if we ride a bicycle at five miles per hour, there will be a certain amount of air resistance to overcome. If we now imagine riding the same bicycle at ten miles per hour, doubling our velocity, then a *linear* assumption would be that the air resistance would also double. It would be proportional to the velocity. In fact this is not the case. The air resistance at ten miles per hour is not merely twice, but eight times, as great. Otherwise said, the air resistance is not proportional to the velocity, but is *nonlinear* and increases as the cube of the velocity. Nonlinearity is not a familiar concept. Perhaps a real world example from biology will help to fix our ideas.

Fish Populations as a Nonlinear System

The *openness to alternatives* and *unpredictability* of nonlinear systems was discovered by Robert May, an Australian biologist, in the 1970s.[5] May was trying mathematically to model—that is, to predict—future fish populations based on a present one using an equation of just two terms: one accounted for reproduction and the second term represented overcrowding due to lack of food.

Let's consider in more detail the question of successive years of fish populations. If there were no lack of food due to overcrowding, the number of fish in a subsequent year would be given by the reproductive factor, r, multiplied by the original population. If, for example, $r = 2$, then the fish population would double each year. We can represent this situation by the following *linear* equation:

$$N_{\text{next}} = rN_{\text{present}}$$

Here ^{N}present is the present population of fish and ^{N}next is the predicted population of the next generation. The equation is linear in that the next generation of fish is proportional to the present generation. Mathematically we would say that the present generation occurs to the first power, and is hence linear, just as a straight line in geometry is represented by first, or linear, powers of its coordinates.

However, if there is a lack of food, then the population will be less than double. That lack for each individual fish can be assumed proportional to the population—say a constant, b, multiplied by the present population. The effect on the population as a whole will depend on the lack for an individual fish multiplied by the number of fish; that is, the lack for an individual fish multiplied by the present population, or $bN_{present} x N_{present}$. Thus, mathematically the overcrowding effect will be a constant multiplied by the square of the fish population, a *nonlinear* effect. This term has a negative sign to indicate that it will decrease the number of fish in the next generation. This gives rise to what is known in biology as the *logistic equation—a nonlinear* equation:

$$N_{next} = rN_{present} - bN^2_{present}$$

The fish population affects itself producing a reduction in successive generations proportional to the square of the fish population—a nonlinear system.

This equation has unexpected behavior that perplexed May when he tried to use it. If the parameter r is less than 1, then the population decreases each year and dies out. If r is greater than 1 but less than 3, the fish population will come to some final value after several generations that is independent of the original population. However, if r is just slightly greater than 3, *there are two possible solutions* for the final population—*bifurcation*.. When r increases to slightly less than 3.5, these solutions in turn bifurcate, so there are now four solutions in a cycle of four years. As r increases further there are eight solutions, and in rapid succession 16, 32, 64, . . . This fish population system has a considerable choice among alternatives, an *openness*. Figure 6.2 illustrates the results given by the logistic equation.

Present-day linear physics is often a good approximation—but we should not confuse such models with reality. Electromagnetic theory and quantum mechanics work amazingly well in many areas. However, these theories are linear in, for example, the electric field of electromagnetic theory or in the wave function of quantum mechanics.

In electromagnetic theory or quantum mechanics the electric field does not produce more electric field and the wave function by itself does not produce more wave function. Another way of expressing linear systems involves the notion of superposition; for example, imagine a certain electric charge producing an electric field at a given point, and when it is absent, a second electric charge produces another electric field at the same point. If electric fields are linear,

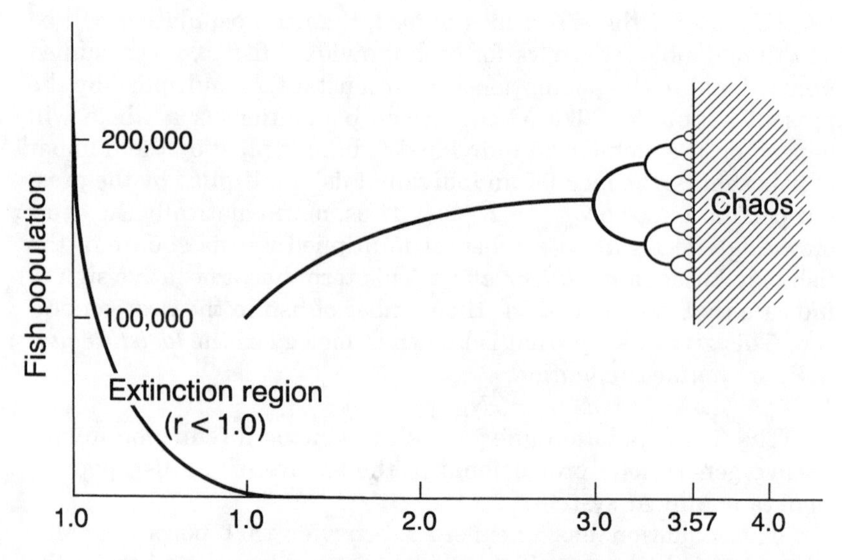

Fig. 6.2. A model of fish population as a nonlinear system

then the combined electric field when both charges are present is the sum of the fields measured independently. If the electric field were nonlinear this would not be the case.

We can also define linear systems as having quantities that enter the equations describing them only as the first power—never, for example, as the square of the wave function or the electric field. We need to recognize that nature is inherently complex and readily exhibits nonlinear phenomena if we look for them. Present-day linear physics, with the exception of the general theory of relativity, is only an approximation, albeit often a good one—but we should not confuse that model with reality.

The Upside-Down Pendulum as an Example of Nonlinearity

An ordinary pendulum makes small to-and-fro movements near its resting point. This motion is periodic and predictable—we use it as a clock. The motion is said to be linear—that is, the force of gravity acting on the pendulum to push it toward its resting (equilibrium) point is proportional to the distance the pendulum is from the resting point (its amplitude). If the pendulum swings twice as far from its equilibrium point, the restoring force of gravity will be twice as great.

However, if we place the pendulum at a considerable angle from its position of rest, the restoring force will not be proportional to its displacement from the equilibrium point—the pendulum system will have become nonlinear. If we increase its displacement further and further, the pendulum bob will eventually reach a vertical position at an angle one hundred eighty degrees from its resting point: It will be an upside-down pendulum.

As shown in figure 6.3, the upside-down pendulum, if released, will oscillate from its position downward and then upward on the other side of vertical and back again indefinitely (neglecting friction). Suppose we give it a small amount of energy and attempt to make it vertical. Even if we position the pendulum bob at this point as carefully as possible, it will in practice be slightly to the left or right of exact verticality.

If we give the pendulum just a very small amount of energy so that it is minutely to the right in the figure, it will rotate instead of oscillating—the motion will change qualitatively. The exact vertical position for the pendulum is again a *bifurcation point*. The system

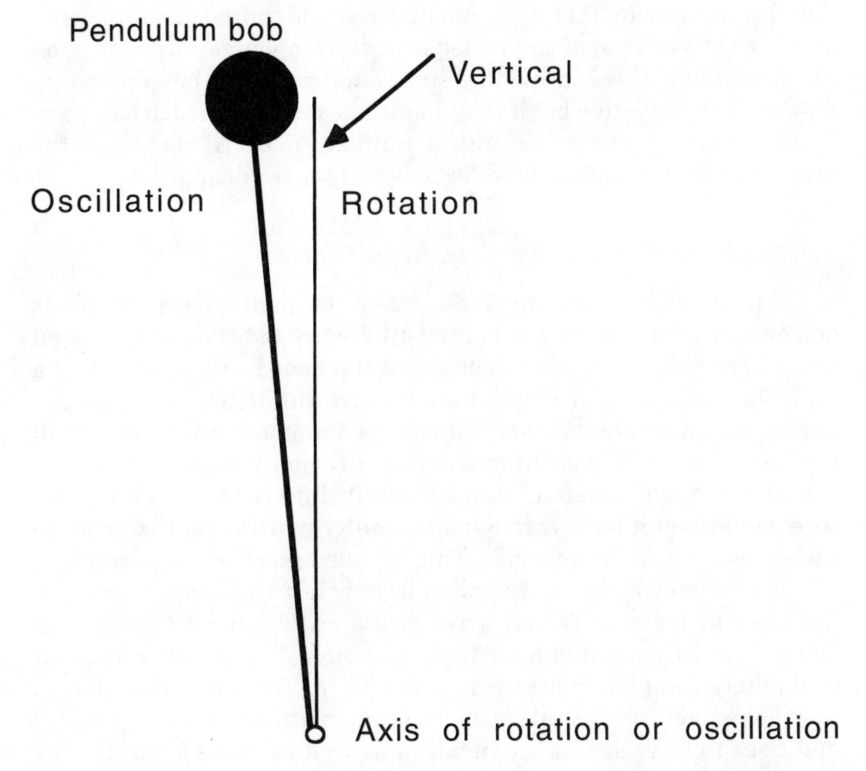

Fig. 6.3. Upside-down pendulum as an example of a nonlinear system near a bifurcation point

selects either of two directions: oscillation or rotation. At this point *the system is unpredictable* in the sense that we cannot in practice specify its position and energy at the vertical point accurately enough to say what its subsequent motion will be. It can select between two alternatives in this case, an openness.

In process terms, a system such as a pendulum or a rock is called an *aggregational society*. Such societies are merely aggregational because they do not have a dominant organizing member. Systems that have the latter are termed *compound individuals*, for example, such individuals might have a nervous system, out of which a mind, constituting the dominant member of the whole society emerges.

In *A Process Philosophy of Religion* Griffin explains: "'. . . the individual actual occasions making up a rock have mentality, which means they have spontaneous functionings. But for lifeless matter these functionings thwart each other, and average out so as to produce a negligible total effect.'[6] The scientific study of such aggregational societies can, therefore, ignore these individual spontaneities; a science of average effect is adequate. In compound individuals, on the other hand, these individual spontaneities are no longer negligible, because they give birth to a dominant member, which can coordinate these spontaneities into a unified effect. A science of the average, based on impartial effects, is no longer adequate."[7]

Another Example of Nonlinearity: Heated Oil

The *creativity* and *interconnection* of nonlinear systems is demonstrated in the case of heated oil. Figure 6.4 shows silicone oil being heated by a hot plate below and the heat being removed by a cold plate above. If a slight temperature difference between the upper and lower plate is maintained, the temperature of the oil will vary in a linear fashion from the warm region below to the colder one above. Such a system is near equilibrium (when the temperature would be uniform throughout), and very little heat is given to the system or removed from it. This is called *thermal conduction*.

If we increase the heat applied from below, the temperature difference will increase. When it reaches a critical point (the system being forced far from equilibrium), the oil will suddenly *create* an exquisitely complex movement, which is far from random. Due to thermal expansion the oil near the lower plate has a lower density and rises to be cooled at the upper plate and then descends. In this process, called *convection*, the entire layer of oil forms into intricate-

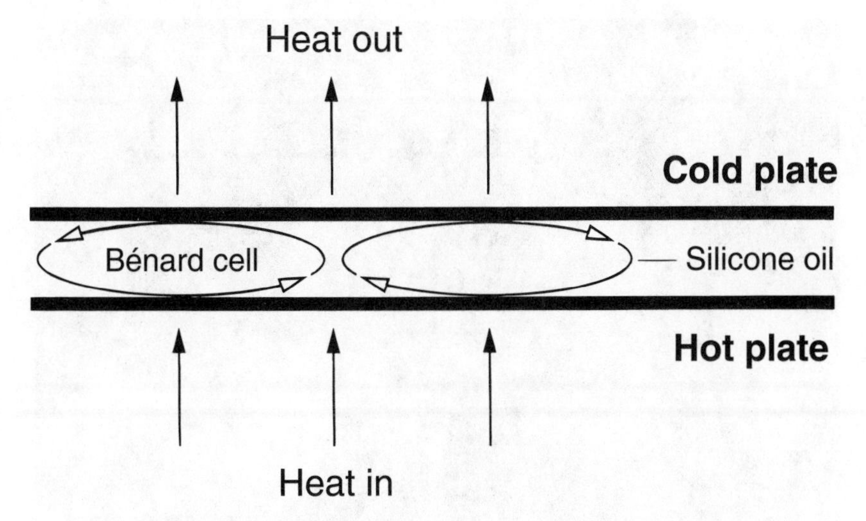

Fig. 6.4. Heated oil as a demonstration of a self-organizing system

ly complex hexagonal patterns of cells, called Bénard cells. Two such cells are shown in figure 6.4. This is a remarkable instance of *spontaneous self-organizing creation*—an intimate *interconnection* among trillions of oil molecules.

The pattern produced is shown in figure 6.5. The hexagonal cells are visible under close inspection in the upper figure; the lower figure shows them more clearly as well as the exquisite structure formed within a cell.[8] Rotation of the liquid is in opposite directions in adjacent cells. At the critical temperature difference between the upper and lower plates when the system changes from conductive to convective flow, the sense of a cell's rotation is established. No matter how carefully the system is controlled, the rotation sense—right- or left-handed—is *unpredictable*. The system is *open* to its own selection among these two alternatives.

A huge number of particles, about a trillion billion in a typical Bénard cell, are organized in a coherent fashion and each of the cells is further organized with respect to all the other cells in the system. Typical intermolecular forces are about a hundred millionth of a centimeter, whereas the cell is millions of times larger. Here we have *self-organization* and *creativity* with complex behavior of a new type: the *dissipative structure*.

The system is dissipative because it requires energy input from the outside environment to sustain it. Such energy is dissipated to keep the organized system functioning. In the case of the convective

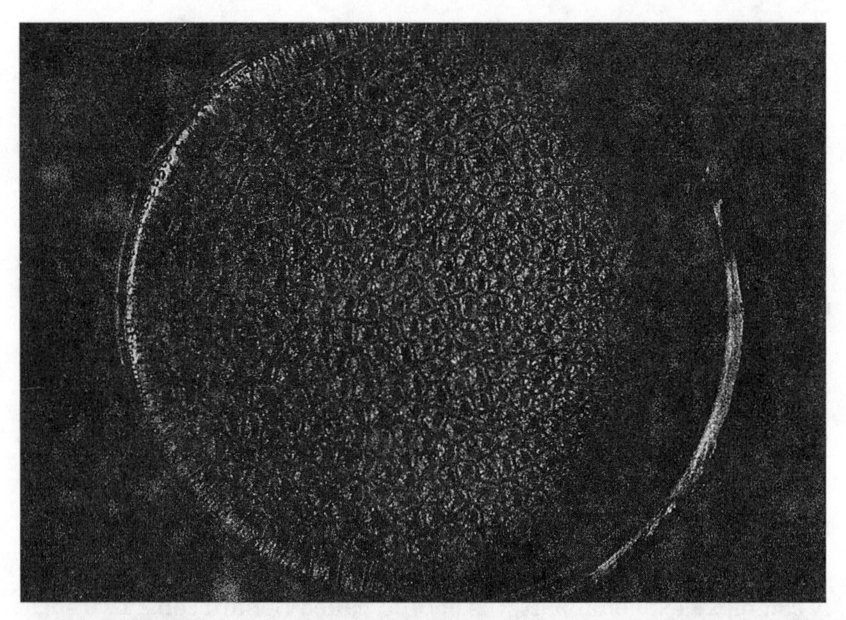

Fig. 6.5. Self-organization of silicone oil molecules. This is an overall view and close-up of the hexagonal convection pattern in a layer of silicone oil one millimeter deep.

oil flow, heat energy is being supplied. Our bodies are examples of such systems. We take in food, water, and air from outside ourselves in order to live. Although the word "dissipation" suggests falling apart, in fact organized structures do arise, as in the case of the heated oil. Dissipative structures are far from equilibrium and are characterized by multiple options, symmetry breaking, and correlations over a macroscopic range.

This behavior is in excellent accord with process thought with its emphasis on creativity and interconnection. In the process view these hexagonal patterns form a complex society of occasions of experience. Through the mutual prehensions of the constitutive events, an active and reactive process, defining characteristics emerge, some of which can be viewed in the figure. This society is maintained as a dissipative structure far from equilibrium by constant energy exchange with its environment.

The heated oil is maintained and forced far from equilibrium by the heat source, which produces a temperature gradient in the oil. A system that is forced far from equilibrium is an irreversible, dissipative system—that is, if the system includes the heat source and the sink, the reservoir where the heat goes after traversing the plates. If we include the heat source and the sink, then entropy will actually have increased overall. However, if we examine only the oil itself when it forms the pattern of hexagonal Bénard cells, we are dealing with a highly ordered, complex, low-entropy system. Far fewer states are available to the oil molecules under these conditions.

More Examples from Biologic Systems

We are beginning to understand some physical dissipative systems, such as the heated oil, but biologic systems are much more complex. One of the characteristics of nonlinear systems is their sensitivity to initial conditions. At a bifurcation point, such as the upside-down pendulum in its highest position, just a very slight change in the position of the pendulum bob will dramatically affect its future behavior.

Peter Coveney, a theoretical physicist, and Roger Highfield, a science writer, point to an example of biologic nonlinearity that has been successfully modeled mathematically. This model describes the rhythms of sugar concentration in cells by which they extract energy for life—a sugar clock. Adenosine triphosphate (ATP) stores the sugars we get from plants, and when we need energy to use, its high-energy chemical bonds are broken: "For appropriate concentrations, the amounts of ATP and adenosine diphosphate (ADP) waltz round a competitive cycle; they vary over a period of a minute, in good agreement with experimental values. Thus, glyolitic rhythms became the first confirmed example of a biological dissipative structure, *a self-organizing pattern* in time."[9] The development of biologic systems is *creative, emergent of new forms*, and *unpredictable*, in agreement with process ideas.

Leslie Orgel and his collaborators at the Salk Institute in California have shown that the nucleic acids DNA and RNA, which code the genes for life, are self-replicating, that is, they catalyze their own production. They are said to be *autocatalytic*.[10] This feature provides the nonlinearity that can give rise to spontaneous self-organization and order, as we have seen. Thus in biologic evolution new structures may arise that make for qualitative innovation. They are not reducible to their antecedents but are a real innovation, real emergences. They give rise to complexity and to new levels of reality within nature with their own appropriate laws of behavior.

According to Joseph Briggs and F. David Peat: "It would have taken nature far longer than the age of the universe to come up with a self-reproducing sequence of amino acids like DNA if the process had been left purely to chance. However, the self-organizing ability of chemical reactions that existed on the early earth suggests that rather than being merely a chance occurrence, the order we call life is a variation on a very old theme."[11]

John B. Cobb Jr. and Charles Birch comment:

> When certain arrangements of carbon, nitrogen and hydrogen atoms, together with a few others, exhibit properties that we recognize by the name of enzyme, and other arrangements of the same atoms result in cells that conduct nerve impulses, we have discovered something new about the nature of those societies of events we call atoms with their remarkably stable structures. When they are organized in these particular ways, the resultant events have characteristics they do not have when this organization is lacking.[12]

These characteristics have been formed through bifurcation points and choices that life systems have made over the ages. Once made, other choices are discarded. The selected ones become amplified and stabilized and are used over and over again as new life forms develop. As a result, human chromosomes and those of chimpanzees are remarkably similar, and the energy provided for our cells involves mitochondria from life forms of the ancient past. The developing human embryo recapitulates and incorporates the history of Earth's life. Nonlinear life systems reveal that time is irreversible.

In accord with Prigogine it can be argued that although physics has been very successful employing linear approximations in the microworld and in mechanics and electromagnetic theory, the *real* world is nonlinear. As such it is irreversible in time, so that time's arrow is universal.

A *real* electron, for example in an atom must feel the influence, minimal though it may be, of electrons in other atoms, free electrons, and other charged particle in the universe, which are themselves changing their quantum states. The electron, having a mass, is also influenced in principle by the gravity of the Sun, planets, and distant stars. All of these influences interact not only with the electron but also with each other. They are therefore nonlinear in their effects. As we have seen, nonlinear systems are extremely sensitive to initial conditions, and they are constantly changing unpredictably and irreversibly due to these myriad influences. It is an impossibility even in principle to reproduce them. Prigogine calls this the infinite entropy barrier. So in this view, even the electron's history is irreversible in the real world.

Stated in process terms, the electron is part of the web of connections to all past events in the universe. The irreversibility of time is an expression of this interconnectivity. Each complex system is part of yet another one, ultimately involving the whole of creation.

As Briggs and Peat put it: "Time thus becomes an expression of the system's holistic interaction, and this interaction extends outward. Every complex system is a changing part of a greater whole, a nesting of larger and larger wholes leading eventually to the most complex dynamical system of all, the system that ultimately encompasses whatever we mean by order and chaos—the universe itself."[13]

Even though a given occasion of experience is connected to previous occasions, according to process thought each society of occasions of experience that is a true individual has its own subjectivity. Associated with this subjectivity are the *subjective aims* of the occasions of that society. These subjective aims lie at the roots of the goals of the enduring individual, for example, obtaining food and reproducing.

Chaos

If we return to the example, of the prediction of fish populations, it is found that when the reproductive rate per generation, r, reaches about 3.57, there are an infinite number of solutions, or alternative choices, and the system is said to be *chaotic*. In this region *there is no predictability whatsoever*. A very slight change in initial population will change the prediction in a given year dramatically. As we further increase r, at about 3.8 the system again becomes predictable. But with increasing r we shall again reach a region that is chaotic, and such regions will alternate with non-chaotic ones.

It is important to distinguish between two types of chaos: deterministic and random. The one just described is said to be deterministic in that solutions are calculable in principle. Near the chaotic regions new self-organization can emerge. On the other hand, for randomly chaotic systems at equilibrium its elements are so thoroughly mixed that no organization exists. It is the world predicted by the second law of thermodynamics of maximum entropy and by "heat death."

In practice, calculations in the chaotic regions of far-from-equilibrium systems are impossible due to the extreme sensitivity of the result to the choice of initial conditions. In fact, it can be shown that for some cases the equation does not compute in the sense that there is no more output information than input information—it merely copies or translates.[14]

In the example of the heated oil, we can imagine more and more heat being applied to the plate below. The system is being forced even further from equilibrium. Before we had a regime in which there was self-organization and creativity, but now eventually bubbles of oil vapor will form and rise and form a chaotic situation. The Bénard cells will be destroyed, and it will be impossible to make predictions about what is happening to a given part of the oil. This example illustrates the idea that creative self-organizing systems often occur near chaos.

In analogy to the examples just discussed, Alfred North Whitehead argues that in process thought it is essential for the social order to evolve and to maximize the intensity of experience of its members. He maintains that such evolution will also take place near chaos: "If there is to be progress beyond limited ideals, the course of history by way of escape must venture along the borders of chaos in its substitution of higher for lower types of order."[15] The French and American revolutions are examples of times of turbulence in society. Old loyalties were replaced with new ones and from the chaos of war new ideas emerged. These revolutions generated new concepts of limitations and responsibilities of government and lifted anew the liberties of individual citizens.

Even systems as classical as the calculation of the orbits of the planets can become chaotic because of minor nonlinear influences. If we ask what the positions of the planets will be ten thousand years from now, the calculation becomes extremely sensitive to their present positions and velocities. There are limits to our knowledge of these quantities. As we extend the calculation further and further in time, the calculations become less and less precise and ultimately become chaotic.

The weather is another example of a system that becomes chaotic if detailed predictions are attempted for more than a few days. The atmosphere is a highly nonlinear, complex system. It has been said that the disturbance of a butterfly's wings in Japan can in principle change the climate in the United States a few days later. In the 1960s with the advent of computers, a program was envisaged by the U.S. Weather Bureau that would predict the weather at least a month in advance. In setting up this program, one of the atmospheric scientists, Edward Lorenz, found that it was so sensitive to weather data that no long-range predictions could be made at all. He had discovered chaos in nonlinear systems.

Process Thought and Complex Systems

In this chapter we have encountered the macroscopic world. In that world, the arrow of time is correlated with an increase in entropy or disorder. Time is an essential element of process thought as a given event becomes an object for prehensions of a future event. Thus, from the process view, time is irreversible, even at the microlevel.

Countervailing the general tendency of the universe toward increased entropy, as specified by the second law of thermodynamics, is the order and decreased entropy produced by complex systems. These systems exhibit spontaneous creativity and unpredictable behavior accompanied by *interconnections* among trillions of atoms. These are all concepts that are tenets of process thought.

We can imagine that over time the creativity of complex systems has permitted life to form on Earth—all supported by our low-entropy energy source, the Sun. Since biologic systems are inherently intricate and with many pathways and connections, it is difficult to construct mathematical models of these systems except in a few cases.

The sensitivity of complex systems to initial conditions often makes it impossible even in principle to predict future behavior for long periods. Complex systems are unpredictable in the sense that *it is impossible to specify initial conditions so that the future of the systems can be determined.* This openness and unpredictability, as we saw it similarly in the Heisenberg uncertainty principle of quantum mechanics, is not because of our ignorance or lack of proper instrumentation, but rather is a fundamental property of nature and is very much in accord with process thought.

Since the world is made up of nonlinear systems that are extremely sensitive to initial conditions, it follows that a minor effect may have enormous consequences. In human terms, an individual at the right place and time can indeed have a major effect on future events.

We share with all life some basic needs: finding food and reproduction. However, humans are an extreme cases of complex biologic systems. In the last few million years our complexity has led to consciousness. For humans, and perhaps also for more highly developed mammals, such as chimpanzees and dolphins, self-consciousness has also evolved. Consciousness and self-consciousness permit a vastly greater creativity and intensity of experience than that, say, of a slime mold. They are sophisticated forms of what Whitehead described generally as "mentality."

In Whitehead's view, when we make a subjective decision based on our goals and input from our past, the process is similar in kind but vastly more complex than that of lower life forms and more complex still than for inanimate matter. Whitehead used the terms *subjective decision* and *subjective aim* in a broad metaphysical sense in relation to much simpler systems than human ones. One of the purposes of such descriptions is to emphasize the generality of process thought.

The word "enjoyment" also has a much broader meaning in process thought. It usually refers to the satisfaction we get from an insight, from a creative experience, or from simply the sense of being alive. However, Whitehead employs the word in the broadest metaphoric sense so that every occasion or event can be said to "enjoy" its own existence. Birds sing and dolphins play for enjoyment. Amoebas and even atoms, so some slight degree, enjoy their existence.

Reductionists claim that all of nature will be explained eventually by the fundamental laws and constituents of matter, understood as insentient bits of mass energy. In my opinion this prediction will not be borne out in complex systems, especially those that process thinkers call compound individuals. It seems to me that at each level of organization new laws apply. Since the whole is nonlinear, interactive, and inherently creative and unpredictable, it is greater in complexity than the sum of its parts. Novel kinds of organization and activity occur. For example, biologic systems are characterized not only by their molecular constituents, but also by their historical accounts of change and cumulative selection as well.

We have seen that systems that are forced far from equilibrium are sometimes capable of impressive *creativity* and *emergence*. These systems often border on the chaotic, as the fish population example demonstrates. A tenet of process thought is that the universe is evolving toward increasing order and complexity. Nonlinear systems and their sometimes chaotic behavior are essential to the ongoing creative process. As in quantum mechanics, in complex systems also we have an interplay between chance and lawful behavior. We find laws of structure and organization of the wholes that cannot be reduced to the some of their parts.

Complexity, such as found in life forms, and in particular in human beings, makes possible vastly enhanced enjoyment in the sense of more creativity, satisfaction, and intensity of experience. This is in accord with process theology, to be discussed in chapter 8, which supposes that the divine is luring the cosmos and the individuals within it to ever greater creativity and enjoyment.

CHAPTER 7

The Macroworld, Part II: Cosmology

We now turn to cosmology, the science of the universe as a whole. Again we shall see ideas fundamental to process thought: *interconnection, an openness and choice among alternatives*, and *creativity*, this time on a grand scale. We shall also appreciate the fact that our very existence appears miraculous. If the physical interactions in the evolution of the universe had been only slightly different, conditions wouldn't have existed for biologic evolution even to begin. We shall see further that the universe is continually evolving—it is qualitatively different now from the way it was billions of years ago. This, too, is in accord with the time-dependent evolutionary ideas of process philosophy.

Both the order and the novelty in the universe are impressive. There is an inherent urge toward greater and greater complexity. According to the present model of the universe most accepted by cosmologists, the Big Bang, a sea of radiation cooled to form quarks, then quarks formed protons and neutrons. These joined to form nuclei and, with further cooling, atoms were created (after about a third of a million years). Atoms formed galaxies and stars, and billions of years later at least one planet was created where life could evolve. The evidence for a creative force in the universe is strong. Although it has not been a smooth evolution—stars explode, life forms come and go—there is a clear trend toward more complexity and capacity for enjoyment, in accord with the process view.

In this chapter we will examine in greater detail this process of cosmic evolution. In the next chapter we shall reflect that the creativity of the universe is possible because it is lawful. The universe is not chaotic or random, but rather appears to be in accordance with the "laws" of physics no matter where or at which times in the distant past we look.

The Big Bang

Today most cosmologists agree that the so-called Big Bang hypothesis is the best explanation we have of the beginning of the cosmos. It is a theoretical construct that agrees with most of our observations. Present evidence indicates that the Big Bang happened about 13 billion years ago. All the energy of the universe is thought to have been localized in an incredibly dense and hot region smaller than a hydrogen atom. It was so hot that only primordial radiation existed; matter could not form until it cooled.

It should be noted that at present a controversy exists among cosmologists as to the age of the universe—when the Big Bang occurred. For some time the dynamics of globular clusters of stars and earlier measurements of the expansion rate of the universe gave concordant answers of about 15 billion years as the age of the universe.

However, recent data indicate that the universe has an age of 13 billion years with a possible error of plus or minus 2 billion years at the 90-percent confidence level.[1] This figure permits agreement with star ages and the expansion rate of the universe. For our purposes we shall use the value of 13 billion years for the age of the universe.

This age could be greater if some present measurements continue to be valid. They indicate that the universal expansion is actually accelerating. In that case, earlier in the universe's evolution galaxies would have been moving apart slower than previously considered; thus they would take longer to reach their present expansion rate. This would have taken a longer time, giving a larger value for the age of the universe. More detail concerning an accelerating universe will be given in the following section on dark matter.

In figure 7.1 we see the current version of the Big Bang hypothesis represented. The figure encompasses seventy orders of magnitude in time. Each interval on the time axis is ten times larger than the interval on its left. The earliest time is 10^{-50} seconds. This is 1 divided by 1 followed by 50 zeros—a very short time indeed (remember 1/10 is written in this notation as 10^{-1}, 1/100 as 10^{-2}, etc.). The next interval is ten times greater: 10^{-40} seconds . . . until it reaches the present time, which is 13 billion years after the Big Bang, or about 4×10^{17} seconds.

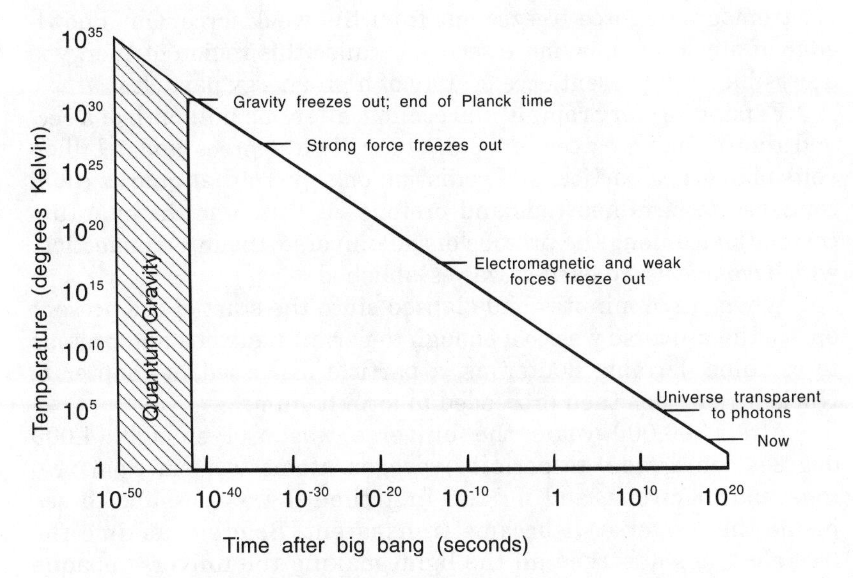

Fig. 7.1. Schematic diagram of the Big Bang

Our present knowledge of physics limits us to consideration of times after the Big Bang that are greater than the inconceivably short time of 10^{-43} seconds, called the *Planck time*.

Before the Planck time the general theory of relativity is not valid because of constraints of quantum mechanics.[2] Thus, the period of the Big Bang before the Planck time is especially speculative since we do not have a theory that unites quantum mechanics and gravity (quantum gravity). The general theory of relativity predicts an expansion of space, which we now observe. According to the Big Bang model this expansion began at the moment of the Big Bang at which time our entire present universe occupied a volume less than that of a hydrogen atom.

The general theory of relativity became valid after 10^{-43} seconds when the gravitational force froze out. Here we follow the unification of forces discussed in chapter 5. An instant later the nuclear strong force froze out and quarks, the constituents of the protons and neutrons, were formed—the first bits of matter. After a trillionth of a second the temperature was 10^{15} degrees and the average energy was 100 billion electron volts, the energy level at which the electromagnetic force freezes out from the weak force. Our knowledge of physics is now more secure because this region of energy is accessible with present accelerators of high-energy particles.

Expanding very rapidly and cooling, after one millionth of a second the fireball was cool enough (10 trillion degrees and 1 billion volts of average particle and radiation energy) so that quarks could condense to form neutrons and protons. At this time the quantum correlations among the protons of the universe, the interconnection, which we discussed earlier, was established.

When three minutes had elapsed since the start of the present epoch, the universe was cool enough to permit neutrons and protons to combine, forming deuterons, a particle discussed in chapter 5. Nuclear reactions then proceeded to form helium.

After 300,000 years the universe was cool enough (4,000 degrees centigrade) to permit hydrogen atoms to form from protons and electrons, and for the first time it was possible to see inside the universe. It became transparent. Before that time the free electrons absorbed all the light, making the universe opaque like a metal. Finally now, after about 13 billion years, the radiation accompanying the Big Bang has cooled to about minus 270 degrees centigrade, about 3 degrees above the absolute zero of temperature. We observe this radiation today as the microwave background, found everywhere in the heavens. We shall discuss this background later.

Evidence Supporting the Big Bang Hypothesis

It may seem arrogant for scientists to say that they know what happened in the first microsecond or less of the universe's existence. What is the evidence for the Big Bang? Why take it seriously? There are four separate sets of observed data that any theory of the cosmos must explain and that the Big Bang hypothesis does explain: (1) the expanding universe, (2) the microwave background, (3) the cosmic abundance of the light elements, and (4) the limit on the number of neutrino families to three. Let's look at each of these in turn.

The Expanding Universe

According to the Big Bang hypothesis, the universe is continually cooling and expanding. This is consistent with observations.

Astronomical measurements show that out to a distance of about a billion light years galaxies are receding from us at velocities proportional to the distance they are from us. This is known as *Hubble's Law*. In the 1920s the American astronomer Edwin Hubble made a series of careful observations of distant stars and established that the more distant stars are from us, the faster they are receding. These and subsequent observations are empirical evidence for the *expanding universe*. The expansion of the universe is also predicted by the general theory of relativity, as was discussed in chapter 2.

The stellar velocities due to the expanding universe are about 22.3 kilometers per second for every million light years of distance. A galaxy that is 10 million light years distant would have a recession velocity of 223 kilometers per second, one that is 100 million light years away would have a recession velocity of 2,230 kilometers per second, and so on. The measurements are made by observing the *redshift* of known spectral lines. Spectral lines are a unique signature for the presence of specific atoms. Since the atoms are moving away from us by the expansion of the universe, their spectral lines are lower in frequency, or have a longer wavelength, than those observed here on Earth—the redshift. The spectra from moving atoms are said to be red shifted. That is, red light has a longer wavelength than blue light so that this effect shifts the wavelength of atoms receding from us slightly toward the red part of the spectrum. Since each atom has its own spectral signature, we also have a means to detect the presence and velocities of other atoms anywhere in the universe.

This is not to say that we are at the center of the expanding universe. Every point in the universe is a center; it is space itself that is expanding. The process is analogous to the role of raisins in making raisin bread. As the bread rises and expands, each raisin becomes more distant from every other, and the raisins that were the most distant from each other initially will have the greatest velocity of separation. It is an omni-centered universe. This observation is in accord with the views of Suzuki, an expositor of Japanese Zen Buddhism, who spoke of two qualities: unimpededness and interpenetration. Unimpededness is appreciating that in all of space each thing and each human being are at the center. Interpenetration means that each one is moving out in all directions penetrating and being penetrated by every other one no matter what the time or what the space.

The Microwave Background

In the early 1960s Penzias and Wilson at Bell Laboratories were setting up a large new sensitive antenna as a microwave receiver for telephone calls from orbiting communication satellites. They were bothered by an insistent microwave background that was always there no matter where in the sky they pointed the antenna. A colleague pointed out to them that two Princeton University astrophysicists, Robert Dicke and Philip Peebles, had accounted for the helium abundance in the cosmos by assuming that the early universe had been very hot. Therefore there would be cooled-down cosmic radiation today, the remnant radiation from the early universe.

So Penzias and Wilson discovered accidentally the remnant of radiation from the Big Bang. We can actually observe the fireball of the Big Bang, which has cooled to almost absolute zero in the thirteen billion years since it occurred. Recently the *Cosmic Background Explorer* (COBE) satellite measured this radiation with high precision. This satellite was launched in 1989 to measure the spectrum and angular distribution of the microwave background over a wide range of frequencies. Spectra from hot bodies were explained by Max Planck in 1900 by invoking the idea of the radiation being emitted in packets of energy, or photons, as we discussed in chapter 3. By using this idea, Planck was able to obtain agreement with experimental spectra from black bodies if the body's temperature were known. The data from the *COBE* satellite were analyzed in the same way.

The satellite was designed to detect deviations from a black body radiation spectrum and was also very sensitive to deviations from *isotropy*, or uniformity of the background radiation in every direction. The results fit the theoretical black body radiation spectrum very well. Investigators found that the data could be fit with only one adjustable parameter, the temperature. That temperature was determined to be 2.735 degrees absolute, or about 270 degrees centigrade below zero—a very cold black body indeed. Figure 7.2 displays these measurements.

The little squares on this graph are *COBE*'s measurements of the brightness of the cosmic microwave background plotted against wavelength. The data falls along a theoretical black body curve for 2.735 degrees absolute (the black line) to a remarkably high degree of accuracy. If the temperature were higher, the theoretical curve would be displaced to the left, if lower, to the right. Thus, the fit shown accurately determines the temperature.

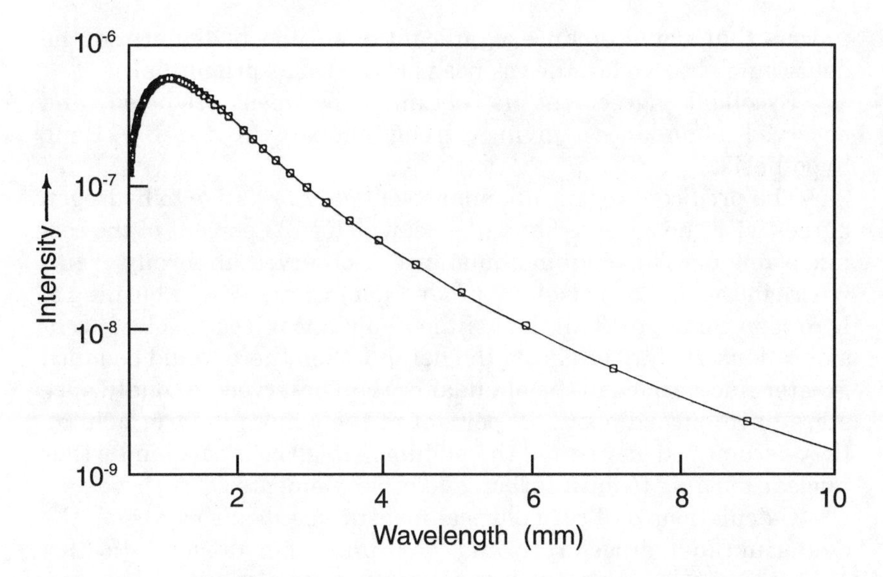

Fig. 7.2. The spectrum of the cosmic microwave background

The wavelength of the radiation is found predominantly in the microwave region, hence its name: microwave background radiation. The black body spectrum is predicted by quantum mechanics and is similar to what we find if we measure the radiation coming from a glowing ember, although in that case the temperature would be much higher than that of the microwave background.

The Cosmic Abundance of the Light Elements

According to the Big Bang hypothesis, a sequence of nuclear reactions occurred after the first three minutes, leading to the formation of helium. First neutrons and protons combined to form deuterons. Then the deuterons combined with other protons to produce an isotope of helium, helium-3. Finally, helium-3 captured a neutron to become helium-4, ordinary helium. The probabilities for these nuclear reactions are well-known from laboratory experiments, so that predictions of the cosmic abundances of hydrogen, deuterium, and helium can be made and compared with the observed primordial abundances. Since there is no known stellar process that would produce significant quantities of deuterium, the deuterium observed in the cosmos is regarded as primordial.

Excellent agreement is obtained between predicted and observed abundances, giving strong support to the Big Bang hypothesis.

The predicted cosmic abundance of helium relative to hydrogen agrees with the observed abundance, which is 23 percent of the cosmic elements. This helium abundance is observed uniformly everywhere in the universe. Helium production in stars via the burning of hydrogen cannot explain the helium abundance, because of two considerations. If stars produced the helium, then there would be much greater fluctuations in the abundance than observed. Secondly, once stars have burned about 12 percent of their mass to form helium, they become red giants and the helium is itself consumed in further nuclear burning to form carbon and other elements.

Calculations and cosmological measurements also agree on the deuterium-to-hydrogen ratio and the lithium-to-hydrogen ratio (2×10^{-5} and 10^{-10}, respectively). No other cosmological theory can explain these data so well. Deuterium is an especially fragile nucleus and is easily destroyed rather than being created within stars. Accounting for the observed deuterium abundance was a major problem before the advent of the Big Bang hypothesis. The primordial production of deuterium as evidence for the Big Bang is discussed by Craig J. Hogan.[3]

The Limited Number of Neutrino Families

The helium abundance is also related to the prediction that there are only three families of neutrinos, which is in agreement with experimental data from particle accelerators (see chap. 5). Malcolm Longair from the Cavendish Laboratory of Cambridge University explains the nucleosynthesis arguments:

> If there were too many neutrino types, the expansion of the universe becomes more rapid resulting in neutrino decoupling occurring at a higher temperature and hence *more* helium being synthesized than observed. Constraining the models to produce the observed primordial helium-4 abundance restricts the number of neutrino species to three, the number known to exist and matching the number of quark flavors.[4]

Problems with the Big Bang Hypothesis

There are several open questions that the Big Bang hypothesis can't answer very satisfactorily, at least thus far. Yet, it is the best explanation we have for the origin of the universe. The problems of flatness and isotropy are explained by another somewhat controversial hypothesis: inflation. One problem has been recently solved. Until the early 1990s astrophysicists were also puzzled over the question of the early structure of the universe, the evolution of galaxies and stars, because the microwave background was known to be extremely uniform. Since matter formed from the radiation sea, the radiation must have been clumped together somewhat in the Big Bang for gravity to aggregate matter into galaxies and stars. However, more precise measurements made in 1992 by the *COBE* satellite showed that there are indeed fluctuations of the order of a few parts in one hundred thousand in the microwave background. This is regarded as strong support for the Big Bang hypothesis.

Flatness

One problem is that the universe exhibits amazing *flatness*. Flatness was explained briefly in chapter 2. The gravity of the mass in the universe opposes its expansion. If the gravity is insufficient, the universe will expand forever. If it is strong enough, the expansion will eventually stop and the universe will contract. If the gravity is just balanced by the expansion, then the expansion will stop after an infinite time, which is the flat universe.

Thus, as explained in chapter 2, depending on the mass of the universe, the general theory of relativity predicts that it may (1) continue expanding so that space would have negative curvature and be saddle shaped, or (2) stop expanding at some time and start contracting so that space would have a positive or spherical curvature, or (3) be on the "knife edge" between the two, a flat plane.

An analogy to illustrate flatness might be helpful. If we imagine shooting a rocket from Earth toward space, then there are again three possibilities: (1) If the rocket doesn't have sufficient velocity, it will eventually descend and fall to Earth again. (2) It may be given enough velocity to easily escape from Earth and go off into space. (3) If we give the rocket just the correct velocity, it will ascend to an altitude above Earth and become a satellite, going around Earth forever (neglecting friction from the remaining atmosphere).

The first case is analogous to the closed universe, in which gravity wins. The second case corresponds to the open universe, in which gravity loses and the universe expands forever, just as the rocket also escapes gravity. The third case corresponds to the flat universe in which gravity neither wins nor loses, but will bring the expansion to a halt after an infinite time. With a nonzero cosmological constant this conclusion is modified even though measurements show that space is "flat" (see the "Recent Measurements" section in this chapter).

To understand flatness we need to know that according to the general theory of relativity, space itself is curved by mass, as discussed in chapter 2. We saw that in the presence of enough mass a light beam would eventually curve back on itself—so in principle you could see the back of your head. This is the situation in a black hole, from which neither light nor anything else can escape (with the exception of Hawking radiation—see chap. 4).

Considering the entire universe, we can ask this question: Is the mass-energy of the universe such that light could eventually escape? If the answer is yes, then the universe would be "open," would have negative curvature, and would continue expanding forever. If the answer is no, the universe, said to be "closed," would have positive curvature. The expansion would stop, and the universe would recollapse into a "Big Crunch."

Observation indicates that neither of these is true, but rather the universe is "flat" delicately balanced between the two. In order for that situation to exist even one second after the Big Bang, the expansion rate initially had to be exquisitely accurate. The accuracy required is one part in 10^{15}—flat, that is, to one part in a thousand trillion.[5] For the universe to be still "flat" today the initial expansion rate would have to be still more accurate, to be almost beyond belief. There is no explanation in the theory of the Big Bang as to this observed delicate balance of the universe between expansion and contraction. Fortunately for us it is so, or otherwise we would not exist!

Isotropy

Another troublesome matter is that the universe has the same composition in every direction as far as we can see for billions of light years, for example, the microwave background is uniform to the order of one part in one hundred thousand across all directions in the sky. The large-scale distribution of stars and galaxies is also remarkably uniform.

The standard Big Bang model shows that radiation sources were as much as ninety horizon distances apart at the time radiation was emitted.[6] (At any time the *horizon distance* is the maximum distance that a light signal could have traveled since our universe began.) In other words, the radiation sources could not possibly be causally connected. How then could all parts of the universe be so similar? This pervading uniformity is called the *isotropy* of the universe.

The idea of *inflation*, discussed in the next sections gives a speculative answer to these questions of isotropy and flatness.

Dark Matter

Introduction

Another embarrassing fact: there is not enough observed matter-radiation (remember mass and energy are equivalent) within our universe to account for the observed expansion rate. This is sometimes called *the dark matter problem*, the important cosmological question of the "missing mass."

Virtually all the hydrogen, deuterium, helium, and lithium observed in the universe has its origin in the Big Bang. Compared to that event, these elements are not formed appreciably by the stellar processes that have created the heavier elements. As just mentioned, by using the Big Bang hypothesis, we can calculate the observed abundances of these elements. Then we can ask how the number of *baryons* (particles, e.g., neutrons and protons that are formed of three quarks and make up what we normally think of as matter) so calculated compares with the number needed for explanation of the expansion of the universe. This comparison reveals the dark matter problem: the calculated number of baryons in the universe is at most 5 percent of the mass necessary to explain either the observed flatness of the universe or its rate of expansion. Most of the matter in the universe seems to be different from our familiar baryonic matter—but we don't know what it is! In any case the total mass, including nonbaryonic mass, is only about one third of that needed to explain the observed flatness of the universe.

Possible explanations for this "dark matter problem" abound, for example, since neutrinos are prevalent in the universe, it could be "closed" if the electron neutrino has a very small mass. Frank Wilczek[7] estimates that a mass of 0.05 eV/c^2 would be sufficient, whereas Phillip J. E. Peebles[8] notes an earlier value of up to 9 eV/c^2. A measurement of neutrinos that arrived along with photons from the supernova 1987 *A* occurred in 1987 (see the next section). This placed a limit of $10eV/c^2$ on the possible mass of the electron neutrino. Limits on the mu and tau neutrino masses are much larger.

However, as we saw in chapter 5, definitive experiments at the Super-Kamiokande neutrino detector in Japan in 1998 showed that the mu neutrino indeed has a mass, perhaps about 0.05 eV/c^2. Under special conditions not yet determined, it could be as high as $3eV/c^2$. In any case the finite neutrino mass will have cosmological consequences not only for the dark matter problem, but in other areas such as the formation of galaxies, since there are about a billion times more neutrinos than neutrons or protons (matter that we can see) in the universe.

It is also necessary to invoke dark matter to explain the motions of stars in galaxies. However, the possible number of neutrinos in a galaxy is limited by the rules of quantum mechanics (Pauli Exclusion Principle). This means that with the limitations of electron neutrino mass just noted, there would be insufficient mass available to explain the stellar motions.

To explain the missing mass, Wilczek[9] discusses several conjectured exotic particles of high-energy physics that haven't been observed. This implies an interesting connection, similar to the idea of neutrinos with a rest mass, linking the very small particles of elementary particle physics with the very large in trying to explain the cosmos.

Another possible explanation comes from accumulating evidence that there really are black holes in the universe. Since light cannot radiate from black holes, they can't be observed directly, but measurements can be made of matter being attracted to massive black holes and emitting x rays and other radiation. The reconfigured Hubble telescope is giving convincing evidence that black holes exist. Small ones could have been produced in the Big Bang, and larger ones as the final phase of stellar evolution (see the next sections). If there are enough black holes, their aggregate mass could account, at least partially, for the missing dark matter.

Recent Measurements

As referred to earlier, the dark matter question took a surprising direction in 1999. Measurements of the light emitted by supernovas that are billions of light years away from us show less light than would be expected if the universe were constantly expanding (for additional information concerning supernovas see the section "Supernovas" later in this chapter).

There are two possible ways to explain these data. First, space could have a negative curvature, be open, not quite flat. In such space, a circle has a circumference greater than $2\pi r$. So there would be more space available for the radiation from an ancient supernova then in flat space, and its light would appear fainter. As a second possibility, it could be that the supernovas are farther away than we expect them to be because the expansion of the universe is accelerating.

Recent precise measurements of the microwave background radiation favor a flat universe, not an open one. Thus, the microwave background measurements seem to favor at this time the second possibility: the expansion of the universe is accelerating. The microwave background is sensitive to the total density of mass-energy within the universe. The rate of expansion depends on the difference between matter, which slows the expansion, and a finite value of the cosmological constant, which can speed it up.

Cosmologists now agree that matter can only account for between 30 percent to 60 percent of the total mass-energy needed to make the universe flat (to agree with the microwave background measurements). Recent articles by the astrophysicists Lawrence M. Krauss, Neta O. Bahcall, Jeremiah P. Ostriker, Saul Perlmutter, and Paul J. Steinbach will give the reader more detail.[10]

A nonzero cosmological constant was first proposed by Einstein when he formulated his general theory of relativity to keep the theory from predicting that the universe is expanding. As mentioned earlier, he later called putting this constant into his theory his greatest blunder after Edwin Hubble later observed that the universe does actually expand.

The value of the cosmological constant needed to explain the observations produces a strange antigravity, a repulsive force, which opposes the attraction of gravity. Gravity acting by itself would slow down the expansion of the universe. This new force explains the observed acceleration of the expansion and its associated energy density could help solve the dark matter problem. The universe expands forever, even though space is flat.

We have no idea about its origin. Furthermore, if the theory is correct, its constant energy density has been ever present since the formation of the universe. It is only now, when the density of matter has been weakened by the dispersal of matter through an ever-increasing volume of the universe, that the force of gravity and the repulsive force provided by the cosmological constant are of comparable size. Why now, one could ask?

The observation of the acceleration of the universe is made possible by a special type of supernova (supernova type 1a) as a "standard candle." These supernovas have been studied extensively and their light output is known to an accuracy of about 12 percent throughout their lifetime. They produce typically a light output of about a billion suns, so that with them one can look back billions of years into the early universe.[11] Computer surveys of the sky are taken with large electronic light detectors on giant telescopes. Then, a few weeks later the same part of the sky is surveyed again. By computer subtraction of the two surveys, a candidate for a new supernova, not seen in the first survey may occasionally be found. Then telescopes around the world verify whether the candidate is indeed a high-redshift supernova. At present a few score have been verified.

As these measurements improve, as well as those of the microwave background radiation, we shall be able to decide definitively whether the universe has negative curvature and is open or is filled with an antigravity energy of unknown origin. At the present time, 2000, the universe seems to be characterized as lightweight, accelerating, and flat. "Lightweight" emphasizes that the matter density is only about one third of that needed to explain the observed flatness of the universe.

Matter-Antimatter Asymmetry

Another problem is: Why is there any matter at all? Or put in another way, why is there an *excess of baryons*, which are the building blocks of matter, compared to antibaryons? In the initial phase of the Big Bang, the quark number was rapidly fluctuating in thermal equilibrium, so there would have been equal numbers of quarks and antiquarks. Therefore, in the initial sea of radiation equal parts of matter and antimatter should have condensed out. When matter and antimatter meet they both are transformed into radiation—thus the Big Bang theory implies that no matter would have been left.

However, a slight asymmetry has been observed that violates to a small extent the idea that there should be equal numbers of particles and antiparticles. All known physical processes generate equal numbers of particles and antiparticles, such as the production of the electron-positron pair, with one exception. The exception is the decay of the neutral *K* meson. The K^o is formed from a down quark and from an antistrange quark. These neutral mesons are observed to decay a fraction of 1 percent more into π^+ mesons rather than their antiparticle π^- mesons. How this asymmetry is related to the actual excess of matter over antimatter in the universe is not clear. That excess is about one part in a billion. Put differently, observations show that there are a billion photons for every baryon. Matter really is an exception in the universe—fortunately for us!

Matter-antimatter asymmetry is in accord with process thought. Asymmetry is one of its fundamental tenets—maintaining as a metaphysical principle that although there is often symmetry in the world, it exists within an overall asymmetry. This is a point that has been especially emphasized by Charles Hartshorne: "This pattern, *symmetry within an overall asymmetery*, we meet again and again. I see in it a paradigm for metaphysics."[12] Space is symmetrical, for example, but time is asymmetrical, leading us from the past to the future, not the reverse. Overall, then, space-time is asymmetrical in the process view.

Initial Low Entropy

How was it arranged that the Big Bang had such a very special initial condition—a highly ordered, low-entropy state—in the midst of the unimaginably hot primordial universe entirely contained within the volume of a hydrogen atom? According to Roger Penrose, the initial state had to be *strictly a radial expansion*, without distorting tidal effects,[13] that is, the expansion had to be absolutely in the outward direction, with no turbulence at all. This allowed the second law of thermodynamics to exist and with it low-entropy sources in galaxies and stars (including our sun), permitting life to be sustained, as explained in chapter 6.

An interesting observation concerning entropy is that if the universe is in fact closed and eventually all matter returns to a "Big Crunch," the result will be a massive black hole. However, a black hole's entropy increases quadratically with the mass it contains. Therefore the entropy would be colossal. It is not at all clear how such a situation could be transformed again into a low-entropy source for a renewed Big Bang expansion.

An Alternate Theory instead of the Big Bang

In the early 1990s, several plasma physicists proposed an alternate explanation of the microwave background radiation and the creation of the very light elements.[14] They argued that early galaxies may have produced an abundance of visible and ultraviolet photons. These would have been absorbed by interstellar dust and reradiated as infrared radiation. Then the microwave background could have been created by further rescattering from plasma filaments that are associated with jets emerging from galactic nuclei. Bombardment of cosmic rays from early stars might account for the small abundance of the light elements. Helium, which is in great abundance, was presumably created in the burning of very heavy stars early in the formation of the universe.

While intriguing, this explanation of the microwave background and formation of the light elements has not gained much favor in the scientific community at this writing.

Answers to the Problems of Flatness and Isotropy: Inflation

Inflation is an idea proposed by the physicist Alan Guth at Stanford University to explain the flatness and isotropy of the universe. He postulates that just after the universe had cooled so that the strong force was frozen out and quarks could be formed (about 10^{-35} seconds after the Big Bang), space suddenly expanded by many orders of magnitude. This special expansion was presumably over by about 10^{-32} seconds, the end of the so-called inflationary epoch. Inflation did not violate the restrictions of relativity, because it was the space between particles that changed—the quarks themselves did not move with velocities exceeding the speed of light. The present observable universe is now expanding into the space created during the inflationary period, which is the "normal" expansion of the universe that we observe today. Figure 7.3 compares the inflation scenario with the standard model proposed by the Big Bang hypothesis. The inflation idea is itself being modified to be in accord with recent developments in particle physics.[15]

Inflation solves the "flatness problem" by making the observable universe flat in the same sense that a small area on Earth's

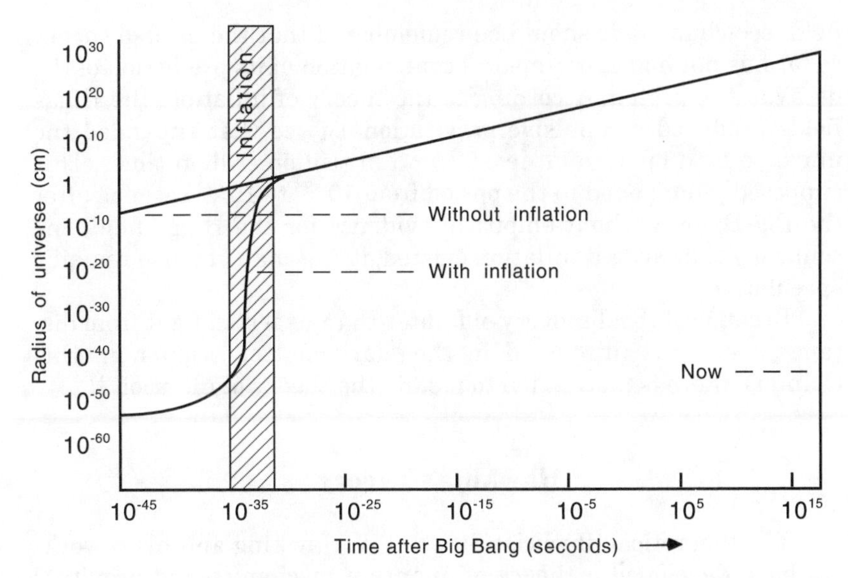

Fig. 7.3. The observable universe with and without inflation

surface appears flat, even though we know it curves with the curvature of Earth. For example, if Earth were only one mile in diameter, a football field would clearly appear to be curved, but we do not observe the curvature because Earth is "inflated" to a diameter of some 8,000 miles. Analogously, we observe space to be flat because in the "inflationary epoch" it was suddenly inflated many orders of magnitude, so that we are unaware of its curvature.

Inflation also explains isotropy. When we observe the microwave background radiation from opposite sides of the known universe, it is coming from distances that are not causally connected, that is, even a light signal could not have traversed the some 26 billion light years of distance during the lifetime of the universe, 13 billion years. Why, then, is the microwave background essentially the same no matter where we look in the sky? Inflation provides the answer by stating that these parts of the universe were causally connected before the inflationary epoch. Therefore, they have a common history even though now, after inflation, they are causally disconnected.

A major problem with the inflation hypothesis is that its cause is only speculative. Guth and another physicist Paul Steinhardt propose a Higgs field to explain it.[16] Higgs fields are named after Peter Higgs, a Scottish physicist. The unified theory predicts this field (see chap. 5). It should be remembered that the unified theory so far has not had experimental confirmation of its prediction of the decay of the proton. According to the theory of inflation, the Higgs fields produced a repulsive gravitational force that expanded the universe by a factor of at least 10^{25}, ten trillion trillion times. This supposedly happened in the period from 10^{-35} to 10^{-32} seconds after the Big Bang. Without empirical evidence for the Higgs field, one could say that even if inflation existed, its mechanism also remains speculative.

Because of the discovery of fainter-than-expected light from distant supernovas (discussed in the "dark matter" section in this chapter), there is motivation to modify the theory of inflation.[17]

Hawking's Theory

The theoretical physicist Stephen W. Hawking and his coworkers have developed a theory of quantum mechanics and gravity.[18] This theory fills in the time before the Planck time, 10^{-43} seconds, after the beginning of the universe. It appears substantially correct but is admittedly not a complete theory, which would involve the incorporation of quantum mechanics into the general theory of relativity. Hawking's theory is perhaps analogous to that of Bohr's model of the atom. It, too, was correct in major ways but was not embedded in the consistent conceptual framework of quantum mechanics. Applying his theory to the beginning of the universe, Hawking suggests that because of quantum mechanical effects, at the beginning there was no time, only space!

Figure 7.4 illustrates Hawking's theory. Here we depict a "*light cone*" of the special theory of relativity for two dimensions of space with the third dimension being the time axis. This cone would be generated from Figure 2.1 of Chapter 2, if we imagine another dimension of space and the figure rotated about the time axis. This cone contains all events in space-time that begin at the origin. For example, if there is initially a point at the origin that is not moving, then the trajectory in space-time is just along the time axis.

If a point starts from the origin with a velocity in the x or y directions, then its space-time trajectory lies within the cone. The trajectory of the massless photon, which travels with velocity c, is on the edge of the cone, depending on the photon's direction. This cone is a substitute for what really should be a four-dimensional figure with three spatial coordinates and the time.

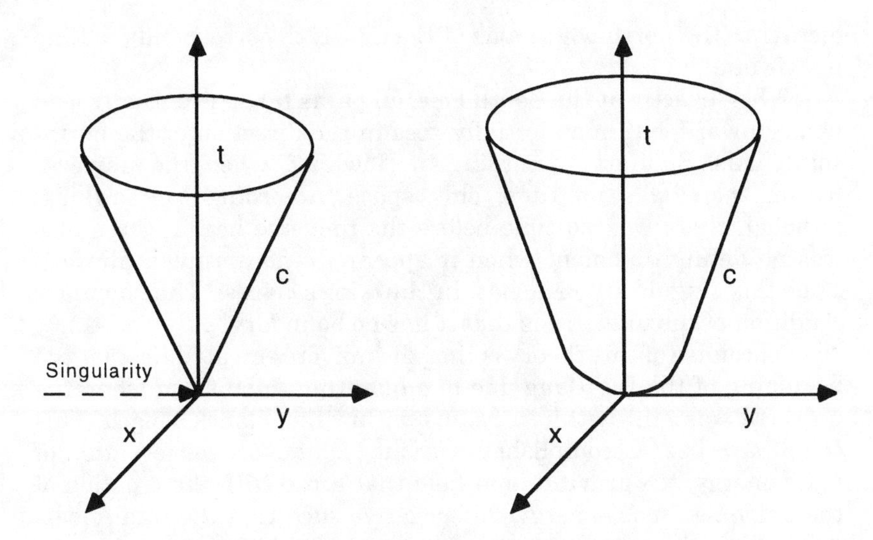

Fig. 7.4. Light cones at the origin of the universe

The beginning of the universe, or $t = 0$ for the Big Bang, is a point at the apex of the cone, and time continues linearly from that beginning. This point is called a *singularity* because here all mathematical description is meaningless and the laws of physics are also not valid.

However, if we invoke quantum mechanics, as Hawking and James Harte have done, the Heisenberg uncertainty principle requires the light cone, instead of coming to a point, to be rounded into something like a hemispheric cap. This is also shown in the figure. The radius of the cap is the *Planck length*, which is very small, about 4×10^{-33} centimeters. This cap forms during the *Planck time* from the beginning of the universe, which is about 10^{-43} seconds (see note 2). While very small, this radius is still infinite compared to a point singularity, and its existence would drastically change our ideas about the origin of the universe.

Above the point where the hemispheric cap joins the cone, time runs forward in the usual way, but below it, for earlier times, time gradually turns into space and as we get finally to the bottom of the cap, it disappears altogether. The situation is similar to being at the South Pole and asking which direction would take a person *directly* toward the North Pole. The answer is that there is no such direction, since all directions at the South Pole are perpendicular to the north-south axis of Earth, which corresponds to time in this analogy.

When exactly at the South Pole, all paths taken eventually lead us northward, but none initially goes in the direction of the north-south axis. Similarly, according to Hawking, when the universe began, there was no time, only space. According to Hawking, although there was no time before the universe began, there also was no definite moment when it appeared—no abrupt beginning. Thus this singularity vanishes. In Hawking's words: "The boundary condition of the universe is that it has no boundary."[19]

A premise of this theory is that the universe created itself at the beginning of the Big Bang due to a quantum fluctuation when the universe was still within a radius within the Planck length. This could have been accomplished without violation of conservation of mass-energy: the gravitational field that arose with the creation of the universe's mass-energy was negative such that the total mass-energy including gravitational energy remained zero.

We have seen that particles are created *ex nihilo* in the vacuum, so that the vacuum is not literally nothing. The assumption here is that quantum mechanics was valid under the awesome conditions of the Big Bang when matter had a density of 10^{94} grams per cubic centimeter. Our actual universe of four-dimensional spacetime is a choice from an infinite number of initial conditions in the "quantum soup." Harte and Hawking chose a particular initial condition because of its mathematical elegance.

Hawking's theory is based on quantum mechanics and hence is time-reversible. As mentioned in chapter 6, Roger Penrose, a theoretical physicist, believes that a proper theory of quantum gravity would remove the time reversibility of quantum mechanics, making it irreversible in accord with the macroworld and with the second law of thermodynamics.[20]

Paul Davies in *The Mind of God* interprets Hawking's result as follows:

> Although Hawking's proposal is for a universe without a definite origin in time, it is also true to say in this theory that the universe has not always existed. Is it therefore correct to say that "the universe created itself"? The way I would rather express it is that the universe of space-time and matter is internally consistent and self-contained. Its existence does not require something outside of it. So does this mean that the existence of the universe itself can be

"explained" scientifically without the need of God? Can we regard the universe as forming a closed system entirely within itself? The answer depends on the meaning of the word "explanation." Given the laws of physics, the universe can, so to speak, take care of itself, including its own creation. But where do the laws of physics come from?[21]

Galaxy and Stellar Formation

We now follow further the spectacular *creativity* of the universe in the thirteen billion years after matter was formed by the Big Bang. All the universe is *interconnected* since all matter, including ourselves, came from the Big Bang. Stars and galaxies were formed and died; exploding supernovas spilled out heavier elements into the cosmos, including our own about six billion years ago. We of the solar system are also *interconnected* because of our common origin in that explosion. We are one with all the plants and animals of Earth—stardust together. We have *no way of predicting* stellar birth and death, but we can observe it back billions of years in time and billions of light years away in space. The evolution of the cosmos is a further example of process thought, which we have seen in previous chapters in the microworld and on a human scale.

Early Evolution of the Universe

The universe has been creative on a grand scale for about thirteen billion years. A billion years after the Big Bang, gravitational attraction formed clouds of hydrogen molecules into galaxies. In order for galaxies to form there must be a small initial concentration of matter, which can then grow as a result of gravitational attraction.

One puzzle, mentioned earlier, has been that the microwave background radiation in the universe is remarkably uniform. Since matter was formed from this initial radiation sea in the early universe, it is difficult to account for the unevenness required for galaxy formation. However, in 1992 painstaking analysis of further measurements with the *COBE* satellite revealed that there are indeed faint perturbations in the uniformity of the microwave background radiation (about one part in a hundred thousand).[22]

After millions of years the galactic clouds contracted into protostars. With the further passing of hundreds of millions of years and further gravitational attraction, the protostars were heated by gravitational compression to temperatures of about one million degrees in their centers. This was sufficient to ignite nuclear reactions that burned hydrogen to form helium, with a great release of energy. Thus, the stars began to shine. A star's lifetime is typically about ten billion years. Our own Sun is about halfway through its stellar cycle and will become a red giant engulfing Earth before it dies as a white dwarf about five billion years from now.

The number of galaxies in the universe is staggering—we know about 100 billion at this time. A typical galaxy may contain 100 billion stars, so the number of stars in the known universe is about 10^{22}. This number is 100 times greater than all the grains of sand in all the beaches of Earth.[23] Recently, very low-density galaxies have been discovered, increasing the number of known galaxies significantly. Because of their low density their evolution is much slower, such that even now they may not have reached maturity since they began after the Big Bang.[24]

In 1999 it was observed that about 5 percent of the brightest stars on our own galaxy, the Milky Way, have planets with masses similar to that of Jupiter. If the measurements were more precise, it is possible that smaller planets, of earthly size, would also be discovered. Thus, the 5-percent figure is a lower limit on planetary abundance, at least in our galaxy. If we assume that other galaxies have a similar number of planets orbiting their stars, then there are 5 percent of 10^{22} planets in the universe. Within this enormous number, the chance of other life forms in the universe becomes almost a certainty. It is a humbling fact. Yet, because of the limitations of the special theory of relativity, we have the possibility of communicating with only the nearest stars in our own galaxy. The closest star, Proxima Centauri, is already 4 light years away.

As we look farther away in distance, we look farther back in time. We can observe objects out to distances of about 10 billion light years, which corresponds to a time when the universe was only 3 billion years old, since the universe has been expanding for 13 billion years. Then there were 1,000 times more of the extremely bright objects we call quasars. (We think that a quasar's prodigious energy, typically that of a trillion suns, is produced by matter falling into enormous black holes.) If we look at closer distances corresponding to later times, we find few quasars. One might say that quasars are the dinosaurs of cosmology—we see the light of those that existed some 8 to 10 billion years ago, but we find none now.

There were also many more strong sources of radio waves at earlier epochs. Thus, we conclude that the universe is in continuous evolution and creativity in addition to what we would expect from stellar evolution alone. The "mix" of cosmological entities itself is dynamic and ever changing. We can appreciate the fact that even on a cosmological scale there is evolution in the universe—an evolution ever leading to more novelty and complexity and eventually leading to the maximization of the enjoyment of sentient beings, in accord with process thought.

Supernovas

It seems probable that all the heavier elements that make up the solar system—the Sun, Earth, and all its creatures, including you and me—were formed in a supernova, an exploding star, about six billion years ago. We are all stardust together, sharing this more recent common origin as well as the Big Bang. Through these origins we are interconnected with all life.

A supernova is a colossal stellar event that is created when a star at least as massive as ten suns has burned up all its hydrogen fuel. Then gravity contracts it, heating it and igniting successive nuclear reactions that burn carbon, oxygen, and silicon to produce iron. Iron has maximum nuclear stability and nuclear reactions cannot transform it to produce energy. At this point, therefore, the star can no longer shine, but gravity is ever stronger and the star continues to collapse with such pressure that protons are condensed to neutrons by the reaction:

$$\text{electron} + \text{proton} \rightarrow \text{neutron} + \text{neutrino}$$

This is the reverse of the usual decay of a free neutron into an electron, a proton, and an antineutrino. As the core reaches the density of an atomic nucleus, so that a thimbleful weighs a billion tons, it suddenly becomes almost incompressible and the matter in the shell of the star, incoming at about 20 percent of the velocity of light, rebounds, fed by the outgoing flood of neutrinos. This creates a shock wave that sends the matter of the collapsing shell into space; a flood of neutrinos from the condensation is also released. Neutrinos from supernova 1987A were detected in the Kamiokande detector in Japan (see chap. 5). In a brief interval of a few hours during the collapse and expansion, a gigantic nuclear furnace has forged the heavier elements up to uranium through the absorption of successive neutrons from the intense neutron sea. The rebounding shock wave now disperses the newly created elements into the cosmos. Eventually by gravitational aggregation some of them will find their way to become part of a new generation of stars.

We can get an idea of the time when the supernova occurred that created us by examining the isotopes of uranium. At present the isotope U-235 is only about 1/140 as abundant as the isotope U-238. (They are both chemically uranium, but U-235 has three fewer neutrons in its nucleus than U-238.) It is reasonable to assume from what we know of nuclear physics that the amounts of U-238 and U-235 simultaneously created in the supernova explosion were roughly equal. Since the half life of U-235 is about 700 million years and that of U-238 is 4.5 billion years, the U-235 will have decayed more rapidly than the U-238. So we can ask, how far back in time do we have to go until the two isotopes have equal abundances? This will be the time of the supernova that created our sun and solar system. Figure 7.5 shows graphically how we can do this and displays the result: 5.8 billion years. This appears reasonable, because the age of Earth has been determined by other means to be about 4.6 billion years, and because we estimate the Sun to be about 5 billion years old.

In 1987 a spectacular supernova (labeled 1987A) was observed in the Magellanic cloud that orbits our galaxy, the Milky Way. Astronomers studied it avidly and much of their data confirmed our ideas about supernovas, for example, neutrinos from the explosion were observed in two separate laboratories. After a few weeks, light

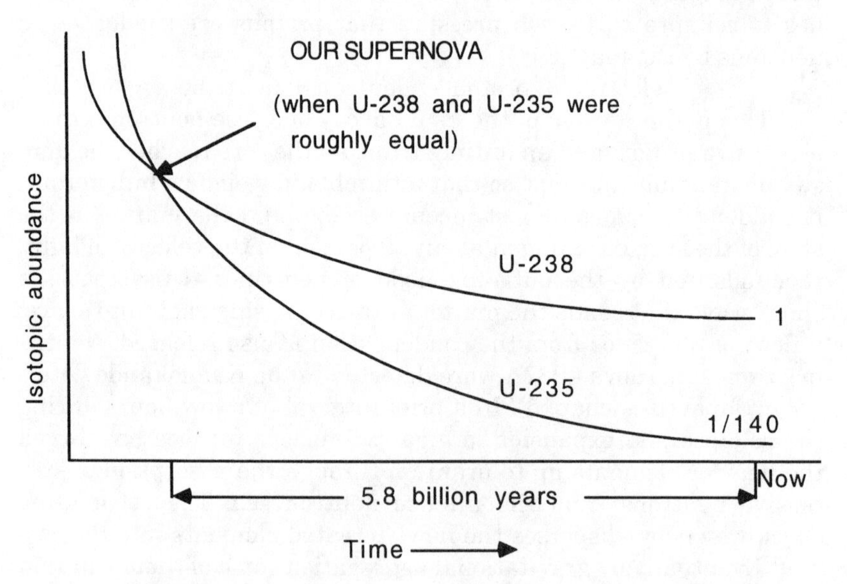

Fig. 7.5. Estimate of the time when our supernova occurred

emission settled into a 77.1-day half life, powered by the radioactive decay of an isotope of cobalt, cobalt-56, which is theoretically predicted to be formed from nuclear reactions in the star's silicon shell. Figure 7.6 displays observations from the Cerro Tolodo Inter-American Observatory in Chile and from the South African Astronomical Observatory.[25] It shows the observed luminosity, or output of visible light, for 250 days following the explosion that formed the supernova. Since the magnitude scale on the left is logarithmic, the radioactive decay of cobalt-56 is linear. The two curves differ slightly because of different assumptions concerning the amount of interstellar dust between Earth and the supernova.

The energy output from 1987A in less than one second was 100 times the energy our Sun will produce in all 10 billion years of its existence. Supernova 1987A actually exploded 160,000 years ago; its neutrinos and photons reached Earth in 1987.

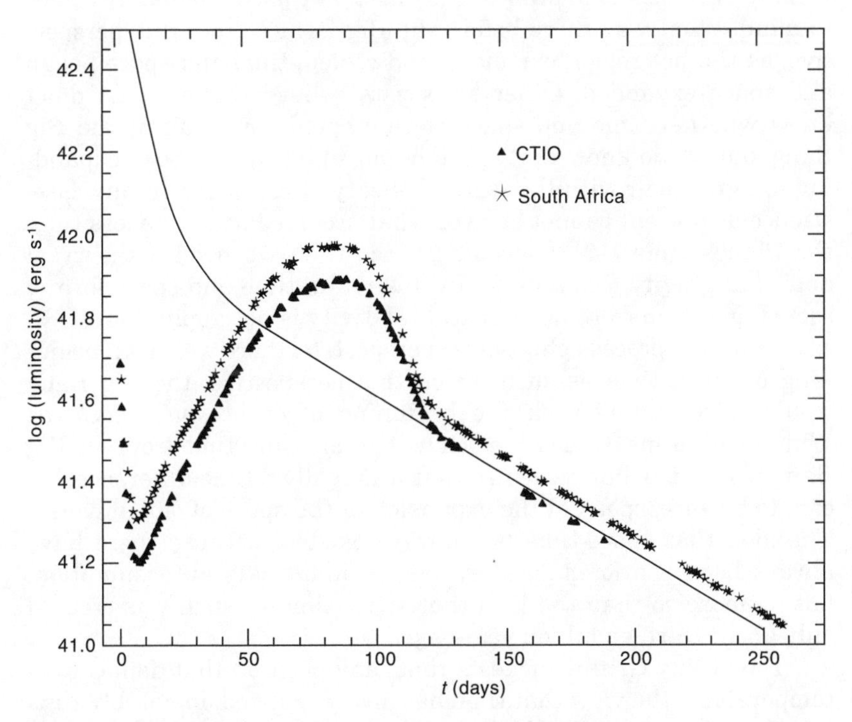

Fig. 7.6. Luminosity of supernova 1987A. The solid curve is a theoretical one based on the radioactive decay of Co^{56}.

Current theories of stellar evolution indicate that if, after a supernova is created, the resulting stellar core has a mass exceeding three solar masses, a *black hole* is formed. If the core mass is less than three but greater than 1.4 solar masses, a *neutron star* (see chap. 2) is formed, if less than 1.4 solar masses, a *white dwarf*. The latter presumably will be the fate of our Sun. A white dwarf shines only because of its remaining thermal energy. When that is exhausted, it solidifies and fades out.

Cosmology and Process Thought

According to some physicists, the Big Bang did not just explode in previously existing space—*it created space*. This is the position of the theoretical physicist Roger Penrose.[26] Since space and time are connected by the theory of relativity, this means that time also would have been created at the moment of the Big Bang. In this view, science has no answer to the question that immediately comes to mind: What was there before the Big Bang? From this perspective, as the hot region expanded and cooled, time and space began and space expanded. Other physicists believe that we just don't know whether time and space were created, or began at the Big Bang, but we do know that space in our universe has been expanding in agreement with the general theory of relativity. In any case, science at present cannot answer what occurred at a time less than the Planck time (10^{-43} seconds), since then we need a theory of quantum gravity (see note 2). In this realm time and space form a sort of quantum froth and general relativity is not applicable.

From the process philosophy perspective, there was not a beginning of space-time as such. From this perspective, the Big Bang would have simply been the beginning of *our cosmic epoch*, as Whitehead terms it. In *our* epoch we can measure time from the Big Bang if we wish, but recognize that it is really a time difference. We can, to be sure, speak of the expansion of the space of *our* universe. The idea that space-time has always existed, because there have always been spatiotemporal events, is admittedly an assumption, but no more so than the idea that space-time as such was created only about thirteen billion years ago.

David Ray Griffin suggests that "any position that denies pantemporalism, the view that time has always existed, inevitably runs into paradoxes, some of which are so strong that they must be called self-contradictions. We can avoid these self-contradictions, if we carry out the logical implications of our premises, only by affirming pantemporalism."[27]

As we discussed early in this chapter, most cosmologists assume that the laws of physics were unified in the first second after the Big Bang. Then the force of gravity "froze out," or separated, followed by the other three forces—the strong force that holds nuclei together, the weak force of radioactive decay, and the electromagnetic force. This was an impressive creative process—producing the physical laws that have been important in forming the order of our present world.

In accord with process thought, the philosopher of religion Ian Barbour argues that history and evolution must be considered when conceptualizing the cosmos: "Uncertainty in quantum physics reflects indeterminacy in the world and not simply the limitations of our knowledge. (Similar contingency is present in the bifurcating of nonequilibrium thermodynamics, random mutation in evolution, and freedom in human life) . . . The cosmos is a unique and irreversible series of events. Our account of it must take a historical form rather than consist of general laws alone."[28]

The cosmology discussed here is just a few decades old. When combined with process thought, it presents us with a new myth, a creation story, appropriate to our times, a comprehensive concept that is in accord with all our knowledge both of the external world and of what we intuitively sense.

Again at a cosmic scale we see illustrations of process thought. The universe is not the same as it was a few billion years ago. It is in a *creative* process of evolution with ever-increasing complexity under a lawful order, starting from a very small intense region of only radiation and evolving to the impressive universe of today. Throughout this evolution there has been a selection among alternatives, an *emergence* of novelty in the processes involved. We are *interconnected* with all life, since the very atoms of our bodies have a common origin in the Big Bang and in the supernova from which our solar system formed. We are stardust together.

CHAPTER 8

Cosmology and Divinity

Our consideration of cosmology in the previous chapter leads naturally to fundamental religious questions. Is there a guiding force in the creation of the cosmos? Why are we here? What is the meaning of our lives, if any? Or is life just a tragedy played out in an indifferent universe? Process theology offers us a framework for considering these questions. It gives us an image of divinity that is consistent with the processes of the universe that we have considered previously. It provides an explanation for the complexity, order, and beauty we see around us and can give meaning to our lives. Process theology is consistent with the metaphysics of process philosophy. We have seen that the latter is in accord with modern science. Thus, process theology provides a means for religion to become more relevant to our scientific-technological society. It offers a spiritual foundation for our modern age.

Sometimes I use words such as "*divine*," "*sacred*," "*deity*," or "*ultimate mystery*" rather than "*God*" because the concept of divinity here is quite different from traditional ones. If I were to use the traditional word, "God," and ask the reader to remember that the word now has different meaning, this might interfere with understanding. As one of my Unitarian-Universalist friends said: "All this God talk is giving me hives."

The Case for Divine Guidance

One of the fundamental mysteries of the awesome creative universe is that it appears to be uniquely constituted so that life could eventually form. What we are discussing here is not

Darwinian adaptation to the environment, but rather a series of fortunate circumstances and physical constants that permitted biologic evolution to begin. There are many such examples in cosmology:

- The universe is amazingly "flat" after thirteen billion years of expansion. This is apparently true even if we must add a mysterious antigravity energy density (see chap. 7). It easily could have been "closed" and have collapsed upon itself billions of years ago or have been "open" with a density of matter too small for galaxies and stars to form. In either case it would have been impossible for life to form on Earth. Human life has occurred only in the last few million years of this long evolution.
- Carbon-12, essential for life since it appears in almost all organic molecules, is created in stars by a nuclear reaction that involves the simultaneous capture of three helium nuclei—a very improbable event. Only the fact that a particular nuclear resonance (a quantum mechanical amplification of the process) is unusually large and in the proper energetic region, makes carbon synthesis possible.
- If the proton-proton nuclear force were a few percent larger, hydrogen nuclei would have formed diprotons or helium-2. Much less hydrogen would have been available, resulting in insufficient fuel for stars to burn.
- If the neutron-proton nuclear force were a few percent larger, much more deuterium would have formed and stars would have burned out long before biologic evolution could have taken place. If this force were slightly weaker, no deuterium would have formed, so that stars couldn't burn at all.
- Carbon is the basis of life as we know it. If the nuclear forces had been slightly stronger, oxygen would have formed in place of carbon. Nuclear forces are barely sufficient for carbon to form. If they were slightly weaker, no carbon would have formed at all.
- If neutrinos had a significant mass, the universe would have collapsed into a "Big Crunch" long ago, before life would have had a chance to develop.
- If neutrinos did not exist, there would be no mechanism

for blasting newly formed elements from a supernova into space so that new stars and planets could form.

- As stated by the theoretical physicist Freeman Dyson, "The weak interaction is millions of times weaker than the nuclear force. It is just weak enough so that the hydrogen in the sun burns at a slow and steady rate. If the weak interaction were much stronger or much weaker, any forms of life dependent on sunlike stars would again be in difficulties."[1]
- Why did the radiation remnant from the Big Bang have the correct, but very small, lack of spatial uniformity (anisotropy) so that galaxies and stars could form?
- Why is there any matter at all? We exist only because there was one part in a billion more matter than antimatter in the early universe. Why is there this fundamental asymmetry?
- Why was the universe formed in an extremely low entropy state, with absolutely no turbulence? This allowed stars such as our Sun to be formed with an entropy increase, conforming to the second law of thermodynamics but still being of sufficiently low entropy to support life on Earth.
- Why is there order in the universe? Why are the laws of physics the same now as in the past and the same everywhere in the observable universe? Starlight reaching us now originated in some cases billions of years ago, yet its characteristics are understandable by the same laws of physics that explain our present physical surroundings.[2] Why is the universe so lawful?

As the theoretical physicist Paul Davies points out,

> The physical world does not merely display arbitrary regularities; it is ordered in a very special manner. The universe is poised interestingly between orderliness (like that of a crystal) and random complexity (as in a chaotic gas). The world is undeniably complex, but its complexity is of an organized variety. The universe has "depth" which was not built into the universe at its origin. It has emerged . . . in a sequence of self-organizing processes that have progressively enriched and complexified the evolving universe. It is easy to imagine a world that, though ordered,

> nevertheless does not possess the right sort of forces or conditions for the emergence of significant depth . . . It is particularly striking how processes that occur on a microscopic scale—say, in nuclear physics—seem to be fine tuned to produce interesting and varied effects on a much larger scale—for example, in astrophysics. Thus we find that the force of gravity combined with the thermodynamical and mechanical properties of hydrogen gas are such as to create large numbers of balls of gas. These balls are large enough to trigger nuclear reactions, but not so large as to collapse rapidly into black holes. In this way stable stars are born.[3]

This list of physical conditions that have allowed us and other life forms to exist is not at all complete. The astrophysicists Brandon Carter, Bernard Carr, and Martin Rees have compiled an extensive list of "fortuitous accidents."[4] Taken together all these special conditions could lead one to the idea that the universe is an extreme form of process—the idea of divine guidance.

The process theologian Norman Pittenger emphasizes the cosmos's support for divinity as follows:

> If self-creativity is the universal principle, if all actualities are partially self-determined or free, what prevents indefinitely great confusion and conflict? Confusion and conflict are indeed real, but they are limited: The cosmos does go on in a reasonably foreseeable way, countless sorts of processes fit together into a varied and beautiful whole, and nobody thinks the universe is likely to blow up of universal conflict. The cosmic order can be viewed in one of two ways: (1) The many self-created creatures harmonize together sufficiently to constitute a cosmos, not thanks to any controlling influence or guidance, but purely spontaneously. Either by sheer luck or their own unimaginable wisdom and goodness, they operate to constitute and maintain a viable cosmos. (2) The many self-created creatures harmonize together to constitute a viable cosmos thanks to some controlling influence or guidance. This influence or guidance can, in a process philosophy, consist only in a supreme form of self-creative power, a supreme form of process which, because of its superiority, exerts an attraction upon all the others, or as Whitehead likes to put it, "persuades", or "lures" them to fol-

> low its directive. I believe a strong case indeed can be made for (2) as against (1) This is the argument from design, or from order, as process philosophy conceives it.[5]

Again, according to Alfred North Whitehead: "Apart from the intervention of God, there could be nothing new in the world, and no order in the world. The course of creation would be a dead level of ineffectiveness, with all balance and intensity progressively excluded by the cross currents of incompatibility."[6]

Given these considerations, it seems a miracle that we exist. For me, they provide impressive evidence for divine guidance in the formation of our universe. In addition to these arguments, there are several others derived from process thought that are beyond the scope of this presentation. They form, in a metaphor given by Charles Hartshorne, strands in a cable in which each strand supports the others. The process theologian David Ray Griffin discusses them in detail in a recent book.[7] These arguments involve the metaphysical order of existence as such, the emergence of novelty, and the world's excessive beauty. He also presents arguments from fundamental aspects of human experience: ideals, truth, the experience of importance, and religious experience.

Process theology incorporates these ideas and provides meaning for life in that it suggests that we are in partnership with the ultimate mystery in creation of the evolving universe. It is also self-consistent and compatible with what science knows about the universe and its evolution.

Some make the argument that the universe evolved randomly and that, accordingly, ours is just the lucky one out of billions of failed universes. But the failed universes are by definition unobservable, so that their existence is clearly speculative. And, assuming that they exist, one can also ask what is their source? Steven Weinberg, the Nobel Prize winner in physics, argues the case against divinity in his book The First Three Minutes: "The more the universe seems comprehensible, the more it seems pointless."[8] For Weinberg we are here by chance and life is tragic.

For me, the view that the universe evolved by chance, while defensible and even courageous, is at its heart unconvincing and unsatisfactory. It is unconvincing because it invokes billions of unobservable universes to try to explain the highly fine-tuned universe that after thirteen billion years of evolution has made our lives possible—to explain the one universe we *can observe*. It is unsatisfactory because it provides no purpose to our being here. Life is a tragedy without meaning. Whatever we create really has no ultimate value. We do not have the excitement of being part of an evolving creation guided by the divine. We have no possibility of a personal relationship with the sacred—the meaningfulness of being copartners with the divine in the creative advance.

More convincing and certainly more satisfying to me is process theology, which is expressed in this excerpt from the poem *Winter Solstice* by Rebecca Parker, president of the Starr King School for the Ministry.

. . . .
Let there be a season
when holiness is heard, and
the splendor of living is revealed.
Stunned to stillness by beauty
we remember who we are and why we are here.
There are inexplicable mysteries.

We are not alone.
In the universe there moves a Wild One
whose gestures alter earth's axis
towards love.
In the immense darkness
everything spins with joy.
The cosmos enfolds us.
We are caught in a web of stars,
cradled in a swaying embrace,
rocked by the holy night,
babes of the universe.

Let this be the time
we wake to life,
like spring wakes, in the moment
of winter solstice.[9]

Process Theology

The study of the divine can be conveniently divided into natural and revealed theology. Natural theology has developed in conformity to the natural world, taking as its authority the observations and experiences that we have in that world. On the other hand, revealed theology's authority is a source that is taken as truth, such as biblical texts. Although process theology is founded on experience and is clearly a natural theology, it has been useful in illuminating many traditional religions. At present there are a number of distinctively traditional theologians who are taking an interest in process theology.

Whitehead was principally concerned with philosophy. But he introduced the divine to make his metaphysics logically consistent and adequate. Theological ideas have subsequently been developed by several generations of process theologians, most notably by Hartshorne. He was not a formal student of Whitehead's but was his postdoctoral student at Harvard in the 1920s. In the period 1925–58 he was a professor in the Philosophy Department and also in the Divinity School of the University of Chicago. There he inspired a new generation of process theologians such as John B. Cobb Jr. and Schubert M. Ogden. Later, he taught at Emory University and at the University of Texas, continuing to publish papers and books well into his nineties. In 1999 Hartshorne celebrated his one hundred second birthday.

Process theology follows naturally from process philosophy. Deity is a necessary part of the metaphysical system, giving it coherence and completeness. In Whitehead's words: "God is not to be treated as an exception to all metaphysical principles, invoked to save their collapse. He is their chief exemplification."[10] Griffin points out that in the process view, God did not create our universe *ex nihilo*, exerting unilateral control, rather: "At each stage, God was working with actual occasions, each of which had the twofold power of creativity: the power to exercise self-determination, then the power to exert efficient causation [influence] on subsequent occasions."[11]

In the beginning of the Big Bang in which our universe began, there wouldn't have been societies of occasions, but, as we saw in the previous chapter, within a fraction of a second simple serially ordered societies, quarks, were formed. Then, upon cooling, nucleons were created, and the process culminated in the formation of atoms, more complex entities, after about a third of a million years of further cooling.

This is just an example, of which we have seen many previously, of ever-increasing complexity in the universe. Process theology offers a possible interpretation of this evolution that gives meaning to our existence. One of its basic tenets is that the divine seeks to increase the enjoyment and creativity, or to enhance the intensity of experience, of all the entities in the universe. This is accomplished through ever-increasing complexity within a reliable order, for example, a human being has much more capacity for intensity of experience than an amoeba. In process theology, the idea of the divine is offered in response to such questions as:

Why is the universe lawful and continually evolving toward novelty and complexity? Why is there not absolute chaos or endless repetition? As we saw in chapter 2, when we look evermore deeply into the universe that surrounds us, we are looking farther and farther back in time, and find the same lawful universe deep in both space and time. And as we saw in chapter 7, the composition of the universe in the past was not the same as it is now. Through aeons of change, the universe has become more complex and self-organized.

About two million years ago, very brief compared to the time for cosmic evolution, the earliest forms of beings that can be called human appeared on Earth. Through us the created universe has become aware of itself. How is this impressive development to be explained? Process theology does so by including the necessity for a sacred component in process metaphysics.

Some ideas of process theology are as follows:

- The divine is an actuality.
- The divine has a primordial nature and a consequent nature.
- The divine lures us toward maximum enjoyment, creativity, and intensity of experience.
- Our connection with the divine is personal.
- The divine actuality employs persuasive wisdom rather than unilateral power.
- The divine knows past and present, and invites us to be coworkers for the future.
- The divine is immanent and transcendent.

The Divine is an Actuality

In process thought there is power in the past. We are continually influenced by our own history and by other entities. The possibilities that affect our decisions also have power. One of Whitehead's major principles is: *Where there is power there must be actuality*. This idea did not originate with Whitehead; he credits it to Plato, saying: "Plato says that it is the *definition* of being that it exert power and be subject to the exertion of power. This means that the essence of being is to be implicated in causal action on other beings."[12] Plato makes the connection between power and actuality clear in a statement in the *Sophist*: "My notion would be that anything which possesses any sort of power to affect another, or to be affected by another, if only for a single moment, however trifling the cause and however slight the effect, has real existence; and I hold that the definition of being is simply power."[13]

Because we are lured to actualize new possibilities in the future, there must be an actual source for future possibilities that reflects the actual world. *In process theology the divine is this actuality*. According to Whitehead, the divine must be actual to account for the power of possibilities. This power is not absolute but one of persuasion—a lure to be our best. In the process view, God is an actuality similar in kind to, but immeasurably greater in degree, than ordinary actualities.

There is a consistency in this metaphysics. There is an actuality for every source of power, past or future.

The Divine Has a Primordial Nature and a Consequent Nature

One of the dilemmas of traditional theology is the invocation of a God that is in all respects necessary and changeless in a world that is changing. By saying that God is necessary we mean that the divine is unique and sufficient unto itself—it needs no cause. Yet our universe is contingent—it could have been otherwise. It is constantly changing. If God is the creator of a contingent world, how can God be unchanging, omnipotent, and omniscient with knowledge of the future? The answer of process theology to this dilemma is to modify the inherited conception of the ultimate mystery, thinking of it as dipolar—as having a *primordial* nature that is unchanging, and a *consequent* nature that becomes more complex and changes with the world. The consequent nature of the divine is affected by the world and affects the world. The primordial nature affects the world too, but is unaffected by it.

In commenting on the *primordial* nature of God, Whitehead states: "Viewed as primordial, he is the unlimited conceptual realization of the absolute wealth of potentiality not before all creation, but with all creation."[14] The forms in the primordial nature lack actuality. They are only potential. And the primordial nature itself lacks actuality, being only an abstraction from the Divine Actuality as a whole. But as such it is the source of the evolutionary process of the universe.

In the process view, there is no "God of the gaps" whereby gaps in the world's causal processes are occasionally filled by divine acts. Rather, the divine acts in the creative decision of every occasion of experience by presenting alternatives according to divine goals. This divine influence explains both the order and the occurrence of novelty in the world. According to Whitehead: "God is the unconditioned actuality of conceptual feeling at the base of things; so that by reason of this primordial actuality, there is an order in the relevance of eternal objects in the process of creation. . . . He is the lure for feeling, the eternal urge of desire."[15]

The primordial character of the deity is eternal in acting as a lure for the evolutionary process and as the primordial ground of order and novelty. Whitehead's term for the divine lure as it becomes a factor in the decision making of an event is the *divine initial aim*. According to process theology, *the goals or aims that an occasion of experience uses as part of its subjective decision are given by the divine initial aim arising from divine primordial possibilities, and tailored in the consequent nature to enhance the intensity of experience of that particular occasion*. The occasion's decision also includes its feelings from past events as well as its subjective goals.

We might be deciding, for example, whether to attend a demonstration in favor of civil rights. Violent acts against demonstrators have occurred in the past. The divine aim may be nudging us to take the risk in the quest for the greater good for the community. The subjective decision of whether to go may or may not be in accord with that divine lure.

Griffin points out how the divine initial aim can account for the "laws" of physics:

> He [Whitehead] also explains, in line with his metaphysical principles, how it can be that the laws of nature are dependent upon God: Worldly occasions all prehend God, thereby receiving initial aims. The Whiteheadian position suggested here is that these *initial aims embody both the metaphysical principles, reflecting God's eternal essence, and the basic contingent laws of this cosmic epoch, reflecting a divine decision at the outset of this epoch*. These divinely derived initial aims are essential to the very existence of the kinds of actualities studied by physics.[16]

Since, in the process view, everything is interrelated, *the divine interacts with the world, and the divine is changed by this interaction*. In this rhythmic way, God is enriched by the world and the world is further enhanced by the sacred lure. The deity interacts with all actual entities, not only with humans. The divine response in the world involves the divine *consequent* nature. We see that by these interactions the divine is not an exception to the process metaphysical framework but the perfect example of it.

Events in the world are harmonized and evaluated in the divine nature. The ultimate mystery is the author of value, the poet of the world—imagining what could be for each of us. What is worthwhile is saved everlastingly in the consequent nature. Whitehead states: "God as well as being primordial is also consequent. Thus, God is the beginning and the end."[17]

The Divine Lures us toward Maximum Enjoyment, Creativity, and Intensity of Experience

The sacred aim of the divine primordial nature lures us toward maximum enjoyment, satisfaction, and creativity. Each such aim is tailored to the individual occasion of experience. We may choose not to heed this call. That is our subjective freedom. So a decision by an occasion of experience, or by a society of occasions, such as a human being, is made not only by feeling past events and reflecting on previously adopted goals but also in experiencing the persuasiveness of the divine with its new possibilities. Since the divine is part of us, feeling our pain and joy, the persuasion is passionate. After the worldly subjective decision is made, whatever is novel and in accord with the primordial nature that guides the creativity of the universe is evaluated and incorporated into the divine consequent nature. It becomes "everlasting." This is a moment-by-moment enterprise.

The divine model as emphasized by Hartshorne, Cobb, and Griffin is that of a serial temporal society of divine occasions of experience rather than a single actual entity, in the same manner as an enduring individual is a series of occasions of experience. At any given moment the divine society is fully actual and interacts with the world. As in worldly occasions of experience, the divine actuality has a physical pole that receives input from the world and a mental pole that harmonizes and evaluates this information—a concrescence that leads to a divine satisfaction and forms initial aims for the world. In the next instant this process is repeated. God is viewed as an everlasting series of temporal occasions of divine experience, each of which has a physical and a mental pole.

Figure 8.1 illustrates this societal view of God and the interaction of the divine with the world.

The lower left part of the figure shows an event, in the sense of an actual occasion, as was illustrated in chapter 1 concerning process philosophy. The event senses, or prehends, past events. New in this figure is the initial aim of the divine in the physical pole. The mental pole harmonizes these data in a process termed *concrescence* along with its own goals, labeled "subjective goals" in the figure. It makes a decision, then reaches a "*satisfaction*." After satisfaction the event becomes "objectively immortal," being available as a datum for future events, including its own future, and for God.

On the lower right is an occasion that is in the process of becoming. Again data from past events, including the prior event in the enduring individual to which it belongs, are being received as well as information from its own goals and a newly tailored initial aim from God. This occasion has not yet made a decision, so no satisfac-

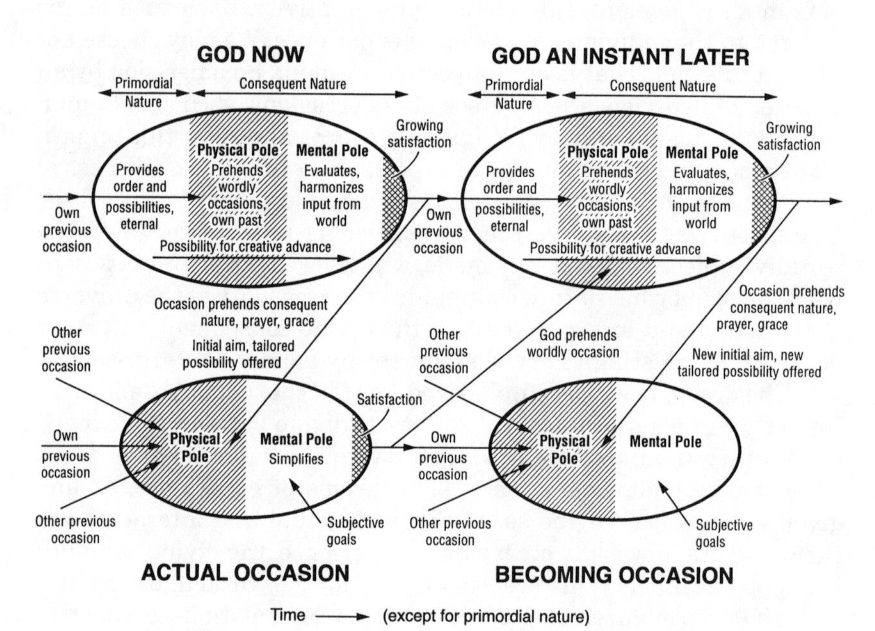

Fig. 8.1. Process view of the interaction of the world and the divine

tion is shown and no datum is available for other occasions or for the divine. Note that time is essential. It flows from left to right in the figure.

As just emphasized, divinity is also an actuality in the process view. To have a consistent metaphysics, it should not be completely different from ordinary actual entities. It is not, although it does have a primordial nature. Positing this exceptional aspect of the divine actuality is necessary in order to explain the order in the universe and all possibilities for its creative advance. This primordial nature is considered eternal and unchanging. It is the source of possibilities for the *divine initial aim* that lures us and all entities to the highest intensity of experience.

On the other hand, the consequent nature of the divine interacts with the world. The figure shows schematically that in the divine mental pole information received from an individual occasion via the divine physical pole is harmonized with the primordial possibilities for the creative advance and is tailored for each individual. It is then available as a new initial aim, as shown for the becoming occasion. Schematically this shows the divine love that is lavished on each worldly event, or society of events, such as a human being.

In the figure, initial aims are shown as arrows directed from the divine satisfaction to the physical poles of the worldly occasions. We can access the divine for prayer and meditation through the same pathway to the consequent nature. Input to the divine physical pole is shown from the satisfaction that has occurred in the occasion on the left. There is no input from the event on the right since it is still in the process of concrescence.

The divine consequent nature assimilates, evaluates, and harmonizes all the data from the world's occasions of experience, and enjoys an increasing satisfaction as the world actualizes its initial aims. The consequent nature is conceived as a series of concrescences and satisfactions. In a continuing process, the divine nature becomes more complex. The conception is of an oscillation between receiving input and providing initial aims, just as worldly occasions prehend the past and then become available as data for future occasions.

Whitehead describes this process in the last chapter of *Process and Reality*:

> The image—and it is but an image—the image under which this operative growth of God's nature is best conceived, is

> that of a tender care that nothing be lost. The consequent nature of God is his judgment on the world. He saves the world as it passes into the immediacy of his own life. It is the judgment of a tenderness which loses nothing that can be saved. It is also the judgment of a wisdom which uses what in the temporal world is mere wreckage.[18]

For Christian process theologians, Jesus was the incarnation of the divine aims here on Earth. As Ian Barbour, professor emeritus of science and religion at Carleton College, states: "Our suggestion, then, is that the ideas of process philosophy may be adapted to the expression of the biblical message in the contemporary world."[19] Many Christian process theologians feel that the divine was able to be effective more fully in certain parts of the past, such as various crucial events in the biblical story. On this basis, they support the integration of Scriptures with process thought.

In the process view, the divine aim, to which we aspire, is to maximize the enjoyment and satisfaction of sentient beings—to maximize the intensity of experience. This encompasses the central Christian idea of love for our fellow humans and extends it to the ecosphere and beyond. Indeed, the process view is deeply ecological. We actually have in us at every moment all previous occasions of experience, including those of plants and animals and even, according to Whitehead, feelings at the molecular level. We are interconnected with the universe.

Again in Buddhist teaching, we individuals are viewed as part of life and we are to be compassionate to all life forms. The Buddhist goal of enlightenment is to rid ourselves and others of *dukka*, or "suffering." Insofar as we attain such a goal (through nonattachment), we and others are able to maximize our enjoyment and satisfaction. Thus process thought is in accord with this fundamental Buddhist precept.

Marjorie H. Suchocki, a Christian process theologian at the Claremont School of Theology, describes how the divine harmonizes possibilities: "Thus as a result of God's own unification of the world with the primordial vision of all possibilities, particular possibilities become relevant to the becoming finite world. These possibilities will reflect, to whatever degree the world can sustain, a transformation of its past in terms of harmony, akin to God's own nature."[20] In this process the divine unifies possibilities in terms of value to promote the maximum enjoyment and satisfaction for the world. This is the divine love for the world. Part of its essence is as evaluator of all possibilities. Thus, process theology is much in accord with the Christian tradition of God's love.

For humans, consciousness is the crown of harmonization because it allows knowledge of interrelatedness and enjoyment to intensify—all creation can be embraced through an intensely conscious existence. Through us, through consciousness, our finite universe becomes aware of itself.

According to process theology, the divine persuasion encourages novelty that can lead to an increased intensity of experience. This process can be envisioned as beauty in a broad sense. As Griffin says in speaking of the soul of the cosmos:

> This cosmic soul has (by hypothesis) an overall aim at beauty in the most general sense of the term, so that those forms of beauty that are possible in a given context are encouraged. This cosmic aim encourages, in Whitehead's phrase, "a three-fold urge: (i) to live, (ii) to live well, (iii) to live better," with the latter meaning "to acquire an increase in satisfaction." If an increased capacity for beauty of experience and thereby positive intrinsic value is generally correlated with increased complexity of structure, the general aim of the soul of the universe would provide a ground for the most pervasive trend evident in the evolutionary process. This explanation would apply at every level, because actualities at every level are occasions of experience originating with prehensions of their environments, which should always include the soul of the whole.[21]

Our Connection with the Divine is Personal

By means of the primordial and consequent nature, we have the possibility of a personal relationship with the divine. From the process perspective, a key characteristic of the divine is love. If we are receptive, God is our advocate.

Through prayer and meditation each of us can prepare ourselves to receive divine grace—to feel, however dimly, the ultimate reality. In the process view, we are constantly receiving lures from the divine. Each lure is tailored for us in an act of divine love—harmonized with our own immediate experience and the divine creative advance to maximize our intensity of experience and enjoyment. In the words of the Quakers, if we seek with a "prepared mind," we can nurture our connection with the divine. Our soul wants to love God, and God too wants to be loved and fulfilled through the connection with us. Sufi teachings tell us to surrender and to fall in love with God—to work with our dark side so that our heart of hearts can be in touch with God's love—and our lives will be transformed.

Palmyre M. F. Oomen, a Whitehead theologian at the University of Nijmegen in the Netherlands, points out that our ability to prehend the consequent nature of the divine, is to "see yourself back through the eyes of someone who loves you, knowing yourself accepted. . . . Prehending God's consequent nature conceptually resembles looking in a mirror, only a mirror in which you see yourself at your best."[22] This can lead to prayer, to a conscience or inner voice, or to a feeling of grace—we are accepted not by what we do, but what we are.

As Suchocki points out, in spite of relationships, there are times when we may feel lonely and empty—life may seem to have little meaning. At such times, if we can be open to the presence of the unseen order, this emptiness is filled with the sense of the divine. We are not alone. This is the aim of the mystic. The divine is always present as our loving companion. We are pushed back into a relationship with the everyday world again, and with our own creativity. Our lives become filled with meaning as we again become copartners with the divine.

Kim Chernin expresses a similar view of spirituality in her book *In My Father's Garden*: "For the time being then, 'spiritual' means for me sensitivity to an unseen order. It means, further, the capacity to take seriously one's relations to this unseen order, so that one can be transformed by it."[23] Again, Martin Buber, the Jewish existentialist, asks us to live in dialogue with the Eternal Thou, to be sensitive to what God may be saying to us in the events of our lives.[24]

There are special times when we may intuitively feel the presence of the divine. My first such experience was during backpacking in the Trinity Alps, as I mentioned before. There was an almost cosmic feeling of being in a sacred garden—of a direct experience of the sacred. It was not knowledge deduced from anything. I just *knew*. The second was at the arrival of our grandchildren. While having our own children we were quite busy surviving, but with grandchildren there is more time for reflection. I was struck with the feeling that although my wife and I did make individual decisions to have children, we were really part of an immense, majestic, ongoing process of evolution of the universe, which we can only dimly perceive. I think these experiences prepared me to be open to the ideas of process theology when I later learned about them.

The Divine Actuality Employs Persuasive Wisdom Rather than Unilateral Power

According to process thought, the divine is not omnipotent but needs the world's cooperation to actualize its possibilities. Our free will is essential for creativity and novelty in this actualization. The divine has power, but it is relational power—the power of persuasion—not unilateral power. So our free will is a fully compatible with such a divinity.

But our free will permits evil to occur. Although the divine lures a society of occasions, such as humans, that is only part of the information to be assimilated. There is also the knowledge of the past and of other possible goals. That is, the society's subjective aims may not be in accord with the initial aim of the divine. The result is that the world, and thereby the divine, are poorer, experiencing the loss of valuable experience that might have occurred. At the human level, we may say that the society of occasions has sinned in being in discord with the divine initial aims.

The process view also might describe sin as acting in such a way as to oppose the sacred purpose in creating the maximum intensity of experience and satisfaction for all occasions of experience, including ourselves and all beings, not just humans. Another way of defining sin is to say that it consists of denying, ignoring, or losing knowledge of the other—of our interrelatedness with the universe. If our desires, the weight of evil, or our life story constrict our freedom to know the other, we sin.

Evil happens not only by an entity's possible interference with or destruction of the capacity of other life forms for enjoyment and creativity, but also by impeding its own. In the process model, since the divine is not all powerful, evil can occur by sheer chance—a part of the riskiness of life. It can occur, and does, in the world. Because of this freedom, creative value and novelty are gained. In this subtle way the divine gives the world unity and purpose in an ongoing evolutionary process and creates in the world ever-increasing complexity and novelty. Evil is part of the process.

The Divine Knows Past and Present and Invites Us to Be Coworkers for the Future

According to process thought, the divine knows the past and the possibilities for the future, but does not know the future. In the process view the divine knows probabilities for the future in ways that we cannot. In some religions, God's omniscience includes knowledge of the future. This may be comforting in the face of the passage of time and inevitable death: God knows all, even the future. But this concept, in hand with divine omnipotence, produces the difficult theological question: Why does God permit suffering in the world? This question does not occur in process theology, because the notion that God "permits" evil implies that God has the power unilaterally to prevent it.

In process thought, divine wisdom replaces divine knowledge of the future. The divine wisdom arising from experience of the world is harmonized with the primordial nature and transformed into the ensuing initial aim.

The consequent nature of the divine feels or has felt every occasion of experience in the world but not those that *will* exist. In the process view, the deity knows it's raining in a particular place, such as San Francisco, but probably not whether it will rain there at a particular time on a particular day next year, because that result is not yet determined, hence unknowable, even by a being who knows everything (then) knowable. This wisdom joins with our own creativity; we become coworkers with the divine. We can share in a small way some of God's qualities. We replace fear with trust and accept, while trying to shape in salutary ways, the unknown future. The divine actuality shapes that future by interacting with the world in unpredictable ways, given the subjectivity of individual decisions.

Process theology is in deep accord with the monumental development of the universe. At the same time it can be intensely personal, as it says that we feel divinity within and about us. We humans have unparalleled capabilities and opportunities for compassion and creativity, but we also have special responsibilities to care for and to further the development of the universe and the beings within it, in partnership with the divine.

These ideas give meaning to our lives. Nothing we do that is worthwhile will be lost, even when five billion years from now Earth is swallowed by the Sun as it becomes a red giant star, or if the universe comes to an equilibrium temperature, a "heat death," according to the second law of thermodynamics.

The Divine is Immanent and Transcendent

The divine persuades events, and this persuasion is taken into account in their subjective decisions. This "friendly persuasion" forms the sacred immanence in the world—what the Quakers term *the Inner Light*, and constitutes a continual and incredible number of complex acts of divine love and compassion—luring us to do our best. The prehensibility of the divine makes access to divinity possible for the worldly seeker through prayer, meditation, and a sense of grace.

Whitehead states, "It is as true to say that the World is immanent in God, as that God is immanent in the World."[25] Aldous Huxley expresses the immanence of the divine as follows:

> Every individual being, from the atom up to the most highly organized of living bodies and the most exalted of finite minds may be thought of as a point where a ray of the primordial Godhead meets one of the differentiated, creaturely emanations of that same Godhead's creative energy. The creature, as creature, may be very far from God, in the sense that it lacks the intelligence to discover the nature of the divine Ground of its being. But the creature in its eternal essence—as the meeting place of creatureliness and primordial Godhead—is one of the infinite number of points where divine Reality is wholly and eternally present.[26]

In the process view, God is involved in our world, but is not of this world. God is transcendent, yet the influence of the divine is pervasive.

Process and Traditional Views of the Divine Compared

In traditional concepts of the divine, God is often perceived as being all powerful and all knowing in the sense of knowing the future as well as the past and present, as noted earlier. If the divine is omnipotent and if the world is full of suffering, then the vision of God that results can be one of aloofness and lack of compassion. It is also sometimes asserted that the misery and injustice in the world is the way God wants it to be as part of the divine eternal plan.

For process theology the divine is, on the other hand, intimately involved with all of us at every moment of our existence. The divine has relational power, the power of persuasion and wisdom, rather than absolute power, so God *cannot* eliminate the evil in the world. In the process view, the possibility of suffering is necessarily involved in fostering enjoyment, creativity, and the intensity of experience. Divine power in process theology is more limited than in some traditional perspectives, acting through wisdom rather than unilateral power, persuading us or luring us to our best through the divine aim. The divine of process theology, in accord with many traditions, is ever present, nurturing us through life. The love of the divine extends to all sentient beings, not just humans.

Although process theology recognizes inevitable limits to the power of deity and its knowledge of the future, it provides for an actuality unlimited in imagination, in affirmation of life, and in fostering creativity. The divine is an ever-present companion who possesses the ultimate in compassion, feeling the pain and sorrow of the world, and the ultimate in response and love for the world. In the divine aim for harmonization and well-being of all creation, the divine is the ultimate in joy, holding all the experiences of the universe together.

To understand fully what the Divine Actuality *is* according to process theology, it is helpful to see what it is *not*. John B. Cobb Jr. and David Ray Griffin have provided a brief list:

> (1) God is not a cosmic moralist, a divine lawgiver and judge who has proclaimed an arbitrary set of moral rules, and who keeps records of offenses. . . . (2) God is not an unchanging and passionless absolute. In this view God is above the cares of the world and is independent of it. . . . (3) God is not a controlling power. This notion suggests that God determines every detail of the world. . . . (4) God is not a sanctioner of the status quo. In this view God has established an unchanging order to the world, and obedience to God is to preserve the status quo. . . . (5) God is not male. The archetype is the dominant, inflexible, unemotional, completely independent (read "strong") male.[27]

Limitations of Process Theology

Although process theology may be attractive because it is in accord with our scientific view of the world, we ought to remember that all such views are subject to change—as is our concept of the ultimate mystery. Process theology is itself part of a process.

As a physicist I am impressed with the lawfulness and order in the microworld of atoms and subnuclear particles described by quantum mechanics. Again in the cosmos we witness enormous creativity and order during the thirteen-billion-year history of the universe—a deep mystery. To me the evolution of order and complexity is strong evidence for a guiding force that is making possible ever-increasing depth of satisfaction and creativity. As exciting as it is, I feel that we should keep a measure of humility because here we face the ultimate reality.

Throughout history we have had *models* of the universe, such as that of the Platonic school that held that the heavens were permanent and unalterable. We have discussed the current model: the Big Bang and the subsequent cosmological evolution. These are models of the universe with a small *u*. They are our feeble approximations to the actual Universe, with a capital *U*, which is not fully knowable. Similarly we have human *models* for the divine in all cultures, and we should remember with Edward Harrison[28] that our models of God are really of god with a small *g*. God with a capital *G* is surely far beyond our grasp, just as is the actual Universe. Still, although we admittedly fall far short, there is a human yearning to search for meaning in our lives, to ask what the universe is, and to seek its source.

Whitehead clearly states the limitation of any philosophy or theology in the last chapter of his major book, Process and Reality:

> In this final discussion we have to ask whether metaphysical principles impose the belief that it is the whole truth. The complexity of the world must be reflected in the answer. It is childish to enter upon thought with the simple-minded question, What is the world made of? The task of reason is to fathom the deeper depths of the many-sidedness of things. We must not expect simple answers to far-reaching questions. However, as our gaze penetrates, there are always heights beyond which block our vision.[29]

The fundamental nature of matter, complex systems, and modern cosmology all fit naturally into the process philosophy of Whitehead[30] and into the process theology he developed along with Hartshorne,[31] and others who envision the divine as a lure, the hidden voice within each of us, the goad toward increasing novelty and complexity, the enticer guiding the creative process of evolution. Finally, we are here with the remarkable ability to comprehend, albeit dimly, the majestic process of evolution in which we are taking part.

Summary of Principal Ideas

Concepts from Cosmology

- We could be the lucky universe out of billions of failed ones.
- Invoking billions of unobserved universes to explain ours is unconvincing and provides no meaning to our lives.
- Our universe seems to be uniquely constituted so that life could form.
- The order in our fine-tuned universe gives evidence for a supreme form of process.

Principal Ideas of Process Theology

- Whitehead developed a philosophy in which the introduction of the divine is a logical consequence of his metaphysics. Where there is power there is actuality. The power of future possibilities rests in the divine actuality.
- The divine provides a *lure* for an event's subjective decision, called *the initial aim*—a persuasion to provide a direction for the occasion of experience, or event, to maximize its depth of satisfaction and intensity of experience. The event, or society of events, doesn't necessarily follow the divine lure, so there is a possibility for evil in the world, but that is the price to be paid for the introduction of novelty and evolutionary advance. Divine love is continually being made manifest for all events, or societies of events, such as humans, in tailoring the initial aim to give that entity the maximum possibility for intensity of experience.
- God saves what is worthwhile from the world, but the divine is neither omnipotent nor omniscient of the future. The divine may know the probability of future possibilities in ways not available to us, but the future is unknown to the divine, since the world's subjective decisions are not knowable before they are made.
- The God of process theology is one of wisdom and persuasion rather than unilateral power.
- The primordial nature of the ultimate mystery provides the guide for *order and creativity in the universe*. This nature is the source of possibilities that the world actualizes.

- Our connection with the divine is personal. The divine interacts with the world through the divine consequent nature, feeling its joy and pain. Through prayer, meditation, and grace we may dimly contact the divine reality. By heeding the initial aim, we become coworkers with the divine in furthering the evolution of the universe. Whatever we do that is worthwhile is saved everlastingly.
- Since process theology is in accord with modern science, it provides a means for religion to become more relevant to our technological age. It provides a spiritual foundation for our modern creation story.

CHAPTER 9

Epilogue: A World in Process

A Look Back

We have now completed our survey of the physical universe, from the world of the extremely small, to nonlinear self-organizing systems on a human scale, and finally to the vast regions of the cosmos itself. At every level we find *interconnection, openness and unpredictability, creativity,* and *increasing order*.

These are fundamental tenets of process thought through its description of events, which are its basic entities. Process philosophy asserts that an event is influenced by, thereby connected to, previous events. The event harmonizes this information with its goals and makes an unpredictable, subjective decision that may lead to creativity and novelty. One of Alfred North Whitehead's principal contributions is the idea that *occasions of experience, or events, have a rudimentary mentality*. We have seen some evidence for this in the quantum domain.

Process thought is appealing to me since it is comprehensive and is compatible with my knowledge of physics. I'm impressed since it agrees with discoveries in physics that occurred decades after its formulation. I like the way process thought leads naturally to what we know about the universe and includes the divine as a logical part of its metaphysics. According to Whitehead, borrowing from Plato, where there is power there is an actuality. The past has the power to affect us and that power comes from actualities, for example, our teachers and parents, who are vast societies of actual entities. Future possibilities also have power, and process thought holds that the actuality behind them is the divine.

Interconnection

Process philosophy teaches us to realize our interconnectedness—that our decisions are affected by past events, and that they in turn will affect future events. We are one. We are all stardust together from the supernova that led to our solar system, or even further back in time, from the Big Bang. We are all connected in the matrix of creation.

A mysterious and indefinitely long interconnection between particles occurs at the quantum level. Quantum mechanics teaches us that when we measure a system, we alter its future. The observer is part of the system—connected to it—not objectively apart as was the assumption of nineteenth-century physics. It is an empirical fact that the microworld is nonlocal and that interconnections are basic. Again, self-organizing systems at the human scale form connections among trillions of molecules, as we have seen in the case of a thin layer of heated oil. We humans are dependent on plants and animals for our survival and depend on each other through our cultures to provide not only the necessities of life, but also the means to expand our creativity.

Within the cosmos there are many interrelationships. Gravity forms galaxies from cosmic dust and with the passage of time, it creates stars. Supernovas spew out elements into space from which new generations of stars, such as our Sun, are born. The gravity of our Sun makes it possible for Earth to be at just the right distance to be hospitable to life.

Interdependence is basic to spirituality. It was long known by Native Americans. They could not understand the idea of land ownership. To them the land was a sacred trust for each to cherish but for all to share. For Native Americans the earth is alive and we are related to all. Buddhists hold that the concept of self is really an illusion of our minds and that humans are necessarily involved in a series of connections from birth to death. Jews and Christians ask: Who is your neighbor?

As we become aware of our connectedness, we become less attached to ourselves and more aware of the mysterious Other of our fellow creatures. We have a feeling of experiencing them more intimately. It is not just an appreciation of the Other, but rather that the Other is in some sense us.

Openness and Unpredictability

Process philosophy asserts that when an individual event makes a decision, that decision has an essential subjective element.

The entity takes into account the past, as well as the goals and possibilities for the future, but the final decision is unique, subjective, and unpredictable. We have seen corroboration of this idea in the physical world.

The Heisenberg uncertainty relations of quantum mechanics tell us that it is impossible in principle to predict the future by calculating the position and velocity of the particles in the universe, as Henri Laplace would have had it long ago. We are restricted in specifying the initial velocity and position so that we can't even begin this theoretical calculation of the future. Nor can we predict the path of an individual particle. We can only specify the probability of its being in a certain position. Neither can we predict when a virtual particle pair creates itself, or when photons and gluons exchange to bind atoms and nuclei. We can only predict the aggregate behavior of thousands of similar particles—that is, the probability that an event will occur.

Again, in nonlinear self-organizing systems we find that often we can never specify their initial conditions accurately enough to permit a long-term prediction of their future course, if at all. We can make predictions of the weather only for a few days at most because we simply cannot define sufficiently all the conditions in the atmosphere at any given moment to enable us to calculate its future behavior.

The cosmos is a dynamic place, for example, the birthing of supernovas. We can observe these events, but we are unable to predict when or where they will happen.

Creation of Order and Novelty

An amazing fact is that the cosmos has created familiar order at times billions of years in the past and to the edges of the known universe. Physical laws seem to be the same even under these extreme conditions of space and time. Why is this so? The answer of process theology is that the divine lure guides the universe toward maximum enjoyment and creativity for all its entities: hence the evolution of the universe toward more complexity and organization. Without lawfulness and order this would have been impossible.

Highly organized life forms, such as humans, are capable of a great deal more enjoyment and creativity than, say, slime molds. Yet, even at that basic level of life there is rhythmic pulsation of creativity, a search for nourishment, reproduction.

The Big Bang was a period of magnificent creativity: our early universe formed in an expanding bubble of intense radiation. As the radiation cooled, elementary particles formed and eventually atoms. Billions of years of creativity followed: galaxies formed, stars ignited. Supernovas produced heavy elements and spewed them into the cosmos. Finally, five billion years ago our solar system formed and conditions for life as we know it were produced on our planet. The universe is not static but continuously evolving—in the past its composition was very different from the present.

Our justified belief in the Big Bang and in subsequent evolution of the cosmos is intimately connected with our knowledge of the nuclear and subnuclear realm. There again we find an order among the subnuclear particles: in the standard model six quarks and six leptons grouped into three families organize the complexity of hundreds of subnuclear particles.

The order we observe in the cosmos gives the strong impression that our existence is indeed miraculous. As I look into a clear night sky, I am easily overcome by a sense of awe. How minuscule I feel compared to its vastness! Yet we are here. I have a sense of belonging.

The parameters of the physical world seem to be fine-tuned so that there was enough time and appropriate conditions for life as we know it to form, as detailed in the previous chapter. In my view, the physical facts that affect cosmology are strong evidence for divine guidance in the evolution of the cosmos.

The Primacy of Events

This table is *really* real. It appears so solid. But is it? In the physical world we have learned that the apparent primacy of substance is due to our lack of ability to perceive the microworld through our senses. Matter is not what we perceive. It is really ruled by events, which are primary in process thought. The mass that we associate with an object is confined to pointlike electrons and quarks that occupy only a trillionth of a quadrillionth of its volume. However, the void that remains is not empty but filled with virtual particle pairs and virtual photons and gluons, a trillion trillion being born and dying every second—a fantastic dance of energy holding matter together. Quantum mechanics teaches us that even the vacuum itself is not empty but filled with the birthing and dying of particle pairs.

Since our memories permit us to recall the past, there must be a record of it somewhere in our bodies and/or our psyches. A record means that there was an event, so that every conscious experience is associated with an event.

Since events are primary, time is fundamental to process thought, as it is in our own experience. Time flows to the future—from birth to death. As we have seen, the second law of thermodynamics agrees with the arrow of time—time flows in the direction of increasing entropy or disorder. Yet out of the disorder, self-organizing systems bring a marvelous complexity and order, such as a human being—what Gregory Bateson termed the *sacramental.*

Science and Metaphysics as Process

Science as Process

In chapter 2 we have seen that science itself is a process. New theories extend old ones as our knowledge of the universe expands. When a superb theory arises, such as the theory of relativity, then this process leads us to a new vision of the cosmos, a transcendence of the old theories. Space and time become a new reality: space-time. Matter and energy form one reality: mass-energy. Time itself depends upon the presence of energetic events as does space. In turn, curved time-space affects mass-energy. We have relationships among these ideas where none existed in nineteenth-century physics.

The general theory of relativity predicts that the universe is expanding, and indeed it is, whereas the Newtonian worldview considered it static. Relativity also predicts waves of gravitational radiation, and evidence for them has recently been established.

Again, quantum mechanics is of more general applicability than classical mechanics. Physicists regard the latter as included in quantum mechanics when quantum numbers are large. Classical mechanics is often a useful approximation of our macroworld, but at the atomic and nuclear levels quantum mechanics is essential. It gives us a description that is strange but compatible with process thought—with events as primary, an openness and unpredictability, and a mysterious interconnection.

More recently, we are beginning to understand complex systems consisting of trillions of atoms with self-organizing nonlinear behavior. New ideas for this new level of complexity are emerging. In 1998 convincing measurements showed that at least one neutrino has mass, the mu neutrino. This upsets the standard model of elementary particles that has been the guide in particle physics for over two decades, for it predicts neutrinos to be massless. So the standard model must now be modified or replaced.

In cosmology, old models of the universe are replaced to explain new observations. The Copernican system replaced the Earth-centered system of the Greeks and of the first millennium A.D. The steady state model of the universe could not explain the microwave background radiation and was eventually superseded by the present Big Bang model.

Science is an ongoing evolutionary process.

Metaphysics as Process

Whitehead's mature formulation of the divine came rather late in his major work, *Process and Reality*. It is probably significant that the divine is not mentioned in the categories (befiting for a mathematician) that he presents early in this book as the fundamental axioms upon which his process philosophy was to be based. In later interviews he admitted that his concept of the divine was somewhat vague in his mind and that conceptual problems existed. David Ray Griffin comments: "Given the fact that Whitehead was a pioneer in working out the doctrine of divine dipolarity and that he came to this task rather late in life, it would not be surprising if his doctrine of God is less clear, and less well integrated into his total position, than other aspects of his philosophy. And this was Whitehead's own judgment."[1]

Whitehead's concept of the divine as a single actual entity led to fundamental metaphysical problems, since enduring individuals in his philosophy are otherwise temporally ordered societies of actual entities. Also in his formulation the intial aim originated solely from the divine primordial nature. Since this nature is eternal and unchanging, it did not provide for initial aims that were tailored to individual occasions of experience so our relationship with the divine was less personal.

In ensuing decades, Charles Hartshorne, John B. Cobb Jr., Griffin, and others have revised the process concept of the divine to be more in harmony with Whitehead's own metaphysical principles. This has resulted in a societal view of God, a temporally ordered series of divine occasions of experience rather than a single actual entity. This was the model presented in the previous chapter.

Any issue of *Process Studies*[2] demonstrates that process metaphysics is a vital, evolving subject, and is clearly in a continual process of change.

A Process Worldview

The chief purpose of this book has been to demonstrate how our knowledge of the physical world—in particular modern physics, nonlinear dynamics, and cosmology—supports the process viewpoint. I narrowed the scope of the book to these connections because these aspects of the physical world are my areas of professional interest. Certainly other areas of knowledge also support the process view. In the biologic realm, for example, ecosystems are interconnected webs of plants and animals. They also contain subjectivity and creativity—all basic elements of process thought.

A proper metaphysics should be comprehensive, to correlate all areas of our knowledge, both of the external world and of what we intuitively sense. I find this to be true of process metaphysics.

We can begin by recognizing the fact that events are primary and substantialism an illusion of our senses. We then can open ourselves to relationships and connections and come into accord with the basic processes of the universe. We are all one and share the common creation, being part of the awesome enterprise of evolution. The future is open, even to the divine, and we are copartners with the divine in creating it. These ideas can give meaning to our lives.

What emerges is a new myth only decades old, a creation story appropriate to our age that gives us a sense of connection and that allows us to place ourselves in the universe—a sense of place that we sorely need.

Process Ethics

Process theology holds that the divine is a creative force in the cosmos that encourages novelty, increasing complexity, and order by means of a persuasive lure. Indeed, after thirteen billion years of evolution, we marvel at the order and creativity at the microlevel, at the human scale, and in the awesome, continuously evolving universe. Because of our own complexity, we humans are best able to use the order in the universe for our creativity.

Process thought teaches us that our purpose in daily life should be to increase the enjoyment and creativity in our own life and in those of others. It declares that to decrease the enjoyment or creativity of ourselves or another entity is a sin against the divine purpose. We not only cause the suffering of that entity or ourselves, but we also deny the divine the possibility of using its creativity. In the process view, when we lose the sense of connection with the Other, we sin.

To me the ethical message is clear: I should treat all beings in our world, not only humans, with loving kindness and compassion—and myself as well. The process view is fundamentally one of connection and nurturing of community. All actualities are intrinsically valuable because they are centers of experience and part of the divine evolutionary process. Indeed, this idea forms the basis for a feeling of responsibility and appreciation toward them. Process theology holds that if we do not conform to the divine lure, evil will result. But our subjectivity is necessary as the ground for novelty and creativity with which we actualize the possibilities of the divine.

If we are committed process thinkers, we are concerned with experience and relationships. Among the life-styles and cultures of our world, we must make value judgments that some life-styles, some aspects of the world's cultures, are immoral while at the same time being mindful that there is a richness to be cherished in all this diversity. To say that it is all relative is not good enough, that is, to indulge in a denial that there are any grounds for criticizing another culture since a sin in one culture may be commendable in another. In a world of overpopulation, nuclear weapons, and global pollution value judgments are imperative.

I have come to realize that through us humans at least, and very recently on a cosmic time scale, the creation has become conscious of itself. With that awareness I feel a responsibility to respect and to nurture all aspects of creation—not only that of humanity but all Earth's ecosystem.

The Environment

In the Western world we have traditionally considered nature as something to be conquered and used for the benefit of humanity or for profit. This was the attitude in the days of the American frontier and is the policy of most of the large international corporations today. Recently, there has been a shift in the attitude of many to the position that we should preserve the environment, or at least parts of it, because it will be useful to humans. At the other extreme of opinion there is a minority viewpoint of deep ecology that posits that we have no more right to be here than any other life form.

In the process view we are connected to all of creation. Each event in an ecological system is made possible by a complex interconnection of antecedent events. So whatever happens in the environment affects us. For me this amounts to an obligation to order my life to enhance the enjoyment of others: the starvation of the peasant, the destruction of fisheries, and the harvesting of giant redwoods, diminishes me.

However, in contrast to deep ecology, in the process view there is in principle a scale of value that gives priority to those entities that have a maximum likelihood of enjoyment and creativity. Establishing such a scale is not an easy task, but process thought provides a theoretical basis, for example, is there a coordinated originality of response that is possible with a central nervous system? Or is it an aggregational society of uncoordinated events, such as a rock?

Plants are best described as aggregational societies but are directly useful to humans and animals as well as having an aesthetic value. We necessarily use other life forms to sustain ourselves, but I try to do this with a constant awareness that there is no sharp separation of myself and others—we are all part of the community of life. I feel a stewardship responsibility to take care of the ecosystem and to use it sparingly and with reverence. For me, besides having an intrinsic value, all entities have instrumental value for the role they play in the ecosystem, and they may also have an aesthetic value.

Environmental concerns are supported by process theology as emphasized by the process philosopher C. Robert Mesle: "We should say that God aims at richness of experience, both for God and the world. God shares the experience of every creature. So the pleasure and pain of bees, bats, and baboons are part of God's life. Their lives matter to God. so if we care for God's life, we should care for theirs."

According to Mesle, "Process theology demands that we stop seeing the happiness of human beings as the sole purpose of God's existence and creative activity. We must respect all of creation and all the other creatures with whom we share this world."[3]

In industrialized nations, separation from nature has resulted in exploitation of Earth's resources with little regard for its ecosystem. Process theology teaches that there is indeed value, that is, the possibility for greater enjoyment, in fostering the most complex growing edge of evolution—we humans. But at the same time it cautions that this not be done if we deem the enjoyment and satisfaction of other species unimportant. Indeed, it teaches that there was such enjoyment and satisfaction long before the appearance of humanity. However, the intensity of experience vastly increased when animals and eventually humans appeared on Earth.

Although process thought has given me a new reverence for the ecosystem, I, as a scientist, particularly need to abandon my tendency toward giving priority to technological solutions to environmental concerns. With such a change in attitude I find my imagination released to help change my way of life and to find less destructive ways of obtaining life's necessities.

Hartshorne stated his environmental view as follows:

> We can try to view man and the rest of nature as one ecosystem in which our species is, so far as possible, complementary, rather than competitive, with the other creatures in the system. To this end we can take zero population growth, or even eventually, and for a time, population decrease, as desirable goals. We can put a burden of proof upon each proposed destruction of wild nature. We can weigh seriously the need for luxuries which use large amounts of energy. . . .[4]

Economics

The earth's resources—fisheries, forests, and food production—are in decline, but the world's population and its expectations increase. In prevailing economic models, which are based on the primacy of substances, only commodities consumed by an individual contribute to that individual's satisfaction or "utility function." Concerns for the satisfaction or sufferings of others are not involved in market activity. Neither are environmental concerns. These are often expressed as "externalities." So the system attempts to remake people to fit into its assumptions: that connections with others and the environment are unimportant. It seems clear to me that the "bottom line" or profit in our economic system should be changed so that it not merely fosters consumerism, but rewards respect for the ecosystem and fostering of community.

Cobb, writing in the concluding chapter of Mesle's Process Theology: A Basic Introduction, demonstrates how present economic theory emphasizes consumption rather than community:

> As economists they subscribe to the principle of "pareto optimality." This principle is that since there is no one who can compare the feelings of loss or gain when these are distributed among different people, we can affirm a net gain only when some are benefited without any loss to others.
>
> This principle has sweeping consequences. It means that when we turn to economists for guidance in public affairs, and when the economists function simply in that capacity,

> we are encouraged not to redistribute wealth but to increase the total amount. There is no sense of either minimal requirements or of sufficiency built into economic theory or the public policies to which it gives rise. An economic theory built on process theism would have very different results.[5]

According to process theology, for example, the divine experiences the misery and suffering that an economic system based on pareto optimality produces. The divine not only experiences that pain, but the pain of experiencing what might have been if the economic system had more compassion, kindness, and a sense of community. If we are copartners with such a divinity, we should take steps to change that economic system.

Race riots in Los Angeles. High iron fences and security systems. Guarded entrances to subdivisions. "We got ours and we will protect it." Those excluded live in ghettos and vent their anger. We witness the high price being paid for elevating individualism at the expense of community. The emphasis on substances as primary—as contrasted with the primacy of events in process thought—has led to individualism as the prevailing philosophy today in industrialized societies. Such an emphasis has little place for valuing of relationships and community. It has often led to a lack of compassion for our fellow humans, for the Earth, and for the creatures that sustain us.

An example of the valuing of individuals at the expense of community is the economic systems of the "developed" nations. These systems have no inherent means for valuing connectedness to the larger community; rather they extol the economic power of individuals and their corporate interests. On the other hand, we can describe the process view as considering the individual *within* community. It empowers individual values through relationships in a community that values both freedom and diversity.

Almost universal communication has made American consumerism the envy of the world. We have 5 percent of the world's population and use 25 percent or more of its energy, often wastefully. The material resources of this planet cannot support this American standard of living for everyone. The almost unrestrained international marketplace, of which we are the leader, is promoting a way of life that is unattainable for all but a very few of the earth's people. It is a way of life that is often spiritually hollow in the quest for more material things.

As Cobb and Herman E. Daly, an economist, point out, the process view argues for adjusting our economic system so that we can foster relationships and community.[6] It argues for us to find ways, for example, by means of tax incentives or subsidies, to encourage the formation of small businesses in small communities and to revitalize the small family farm. These elements of our society have traditionally been oriented toward human relationships and respect for their environment.

Recent tendencies have been in the opposite direction: foreclosures on farms that families have owned for generations; and the decay of towns, especially in the Midwest, accompanied by the ever-rising influence of faceless international companies and ever greater concentration of wealth. WTO and NAFTA have little to offer to the sense of community either in the United States or abroad. Indeed, their support of large industrial and agricultural enterprises comes at the expense of small communities both here and abroad, since the economic system does not value them.

A Last Word

I feel called to responsible action. I feel a need to find creative ways not only to survive, but also to foster a meaningful life in community. It seems to me that if we have the will and approach the ecosystem with respect and humility, we can even flourish on an earth with limited resources and with other life forms that are finite. I see the future as open and with real dangers, but I don't regard our self-destruction as inevitable.

We have seen in a previous chapter that the world is an interrelated nonlinear system. As such it can at times be very sensitive to the forces within it. At certain propitious times especially, one individual can make a difference. These times call for action. Action apart from the programs of the industrialized nations that mostly meet present perceived needs without regard to future needs or to the ecosystem. These times also call for discarding the habitual detached viewpoint of the East, which can be destructive as well. Process thought invites us to consider the world as an interrelated unity: nature participates in us and we in it. It asks us to share with reverence in the divine adventure of the universe.

Notes

Chapter 1

1. Alfred North Whitehead, *Adventures of Ideas* (New York: Macmillan, Free Press, 1967), p. 128.

2. David Ray Griffin, *A Process Philosophy of Religion* (forthcoming).

3. Whitehead, *Science and the Modern World* (New York: Macmillan, 1925); Whitehead, *Process and Reality* (1929; reprint, New York: Free Press, 1978).

4. The idea that substance is not what it seems is not only validated scientifically, but also is in accord with the Hindu concept of *maya*—that the world we take as real is actually an illusion.

5. Whitehead, *Process and Reality*, p. 79.

6. J. Adler, and W. Tse, "Decision-Making in Bacteria," *Science* 184 (21 June 1974): pp. 1292–94.

7. Griffin, *Religion and Scientific Naturalism: Overcoming the Conflicts* (Albany: SUNY Press, 2000).

8. Ivor Leclerc, *The Philosophy of Nature* (Washington DC: Catholic University of America Press, 1986), p. 159; emphasis added.

9. W. Rahula, *What the Buddha Taught* (New York: Grove, 1959), p. 53.

10. Conrad H. Waddington, *Behind Appearance* (Cambridge: MIT Press, 1970), p. 114.

11. Jack Forbes, "*Kinship Is the Basic Principle of Philosophy*." poem by Jack Forbes, University of California at Davis (reprinted by permission of the author)

12. Whitehead, *Process and Reality*, p. 145.

13. Waddington, *Behind Appearance*, p. 114.

14. R. Kane and S. H. Phillips, eds., *Hartshorne, Process Philosophy and Theology* (Albany: SUNY Press, 1989), p. 7.

Chapter 2

1. K. H. Wolfgang Panofsky and Melba Phillips, *Classical Electricity and Magnetism* (Reading, MA: Addison-Wesley, 1955), p. 233.

2. Milic Capek, "Time-Space Rather than Space-time," in *The New Aspects of Time: Its Continuity and Novelties* (Dordrecht, Netherlands: Kluwer Academic Publishers, 1991), p. 324.

3. For those mathematically inclined, $\gamma = 1/\sqrt{(1 - v^2/c^2)}$, where v is the velocity of the moving clock.

4. There is at present a controversy among cosmologists as to the age of the universe. The accepted value had been 15 billion years, but that is now being challenged. We shall use 13 billion years. For more detail, see chapter 7.

5. A. Eckhart and R. Genzel, *Mon. Not. R. Astron. Soc.* 284, no. 576 (1997).

6. Steven W. Hawking, *A Brief History of Time* (New York: Bantam Books, 1988), p. 104.

7. Yervant Terzian and Elizabeth M. Bilson, eds., *Cosmology and Astrophysics* (Ithaca, NY: Cornell University Press, 1982), p. 90.

8. D. Kleppner, "The Gem of General Relativity" *Physics Today* (April 1993): 9.

9. Ian Barbour, *Religion in an Age of Science* (San Francisco: Harper San Francisco, 1990), p. 111.

10. Whitehead, *Process and Reality*, p. 61.

Chapter 3

1. An electron-volt, eV, is the energy an electron would get if it fell through 1 volt of electrical potential. An electron traveling from the negative terminal of your 12-volt car battery to the positive terminal would have an energy of 12 electron volts. It is a very small unit of energy, for example, if an aspirin tablet drops from your hand a couple of inches to a table, it will

have an energy of motion (kinetic energy) of about three thousand trillion electron volts. One million electron-volts is written 1 MeV.

2. A few more details about annihilation quanta:

When a positron stops in matter, it will be attracted to an electron since they have opposite electric charges. They briefly form a hydrogenlike atom, *positronium*. In less than a millionth of a second, positronium disintegrates into two *annihilation quanta*—photons of unique energy, each of 0.511 *MeV*, which is the rest mass of the electron or positron. These quanta go in exactly opposite directions from the annihilation site in order to conserve momentum. As we shall emphasize later, each elementary particle in physics has its antiparticle. In this case, the positron is the antiparticle of the electron. When a particle and its antiparticle combine, the particles disappear or *dematerialize* into radiation (or in some cases other particles as well). This is an example of the equivalence of matter and energy (via $E = mc^2$) in which radiant energy is created by the disappearance of matter.

3. John B. Jr. and Charles Birch, *The Liberation of Life* (Cambridge: Cambridge University Press, 1981), p. 66.

Chapter 4

1. P. G. Medi, G. F. Missiroli, and G. Pozzi, "On the Statistical Aspects of Electron Interference Phenomena," *American Journal of Physics* 44 (1976): 306.

2. Rita N. Brock, *Journeys by Heart* (New York: Crossroad Publishing, 1988), p. 43.

3. Werner Heisenberg, *Physics and Philosophy* (New York: Harper, 1958), p. 125.

4. Henry P. Stapp, *Mind, Matter, and Quantum Mechanics* (Berlin: Springer Verlag, 1993), p. 127.

5. Fritzof Capra, *The Tao of Physics* (Berkeley, CA: Shambala Press, 1975), p. 213.

6. A. Einstein, B. Podolsky, and N. Rosen, "Can Quantum-Mechanical Description of Physical Reality Be Considered Complete?" The *Physical Review* (15 May 1935): 47.

7. Stuart J. Freedman and John F. Clauser, "Experimental Test of Local Hidden Variable Theories." *Physical Review Letters*, no. 14 (April 1972): 28.

8. A. Aspect, P. Grangier, and G. Roger, "Experimental Tests of Realistic Local Theories via Bell's Theorem," *Physical Review Letters*, no. 7 (August 1981): 17.

9. A. Watson, "Quantum Spookiness Wins, Einstein Loses in Photon Test," *Science* (25 July 1997): 481.

10. D. Bouwmeester, et al, "Experimental Quantum Teleportation," *Nature* (11 December 1997). 575.

11. Stephen Hawking, *A Brief History of Time* (New York: Bantam Books, 1988), p. 136.

12. Richard P. Feynman, *QED: The Strange Theory of Light and Matter* (Princeton, NJ: Princeton University Press, 1985); Feynman, *Quantum Electrodynamics* (New York: W. A. Benjamin, 1961).

13. Alfred North Whitehead, *Process and Reality* (New York: Macmillan, 1929), p. 348.

14. Conrad H. Waddington, *Behind Appearance* (Boston: MIT Press, 1970), p. 116.

Chapter 5

1. David Ray Griffin, "Hartshorne's Process Philosophy," in *Hartshorne, Process Philosophy and Theology* (Albany: SUNY Press, 1989), p. 11.

2. B. E. Meland, *Man and Society: Evolution and Imagery of Religious Thought from Darwin to Whitehead*, in *The Realities of Faith* (New York: Oxford University Press, 1962), p. 41.

3. Frank Close, *The Quark Structure of Matter*, in *The New Physics*, Paul Davies, ed. (New York: Cambridge University Press, 1989), p. 396.

4. D. Normile, "Weighing In on Neutrino Mass," *Science* 280 (12 June 1998): 1689–90.

5. The color quantum number was introduced because baryons such as the Δ^{++} were discovered. This particle consists of three up quarks in what appears to be the same quantum state. This situation violates the *Pauli exclusion principle* that states that particles such as electrons and quarks that contain half integral units of angular momentum, $\hbar/2$ (called *fermions*), can never occupy the same quantum state. By assigning a new quantum number, "color," the Pauli principle is preserved. In fact, three different colors have to be assigned to the Δ^{++}, or otherwise the Pauli principle would again be violated. These colors are often named red, blue, and green. Gluons come in varieties such as red-green, green-red, red-blue, and red-red. When a red-red, blue-blue, or green-green gluon interacts with a quark, the quark does not change color, but changes its direction. There six other gluons that do change the quark color. For arcane reasons, one of these types of gluons is unnecessary, leading to eight different kinds.

Chapter 6

1. Conrad H. Waddington, *Behind Appearance* (Cambridge: MIT Press, 1970), p. 118.

2. Roger Penrose, *The Emperor's New Mind* (New York: Oxford University Press, 1989), p. 351.

3. P. Weiss, "Time Proves Not Reversible at Deepest Level," *Science News* 154 (31 October, 1998): 11; Alexander Hellemans, "Italy's KLOE Sets Sights on CP Violation," *Science* 284 (23 April 1999): 568.

4. David Ray Griffin, "Pantemporalism and Panexperientialism," in *The Textures of Time*, Paul A. Harris, ed. (Ann Arbor: University of Michigan Press, 1999), p. 32.

5. James Gleick, *Chaos* (New York: Penguin Books, 1988), p. 69.

6. Whitehead, *Adventures In Ideas* p. 207.

7. Griffin, *Process Philosophy of Religion*, p. 36.

8. Miguel Contreras, Instituto Hidráulico, Madrid, Spain

9. Peter Coveney and Roger Highfield, *The Arrow of Time* (New York: Fawcett Columbine, 1990), p. 225

10. Ibid., p. 217; Joseph Briggs and F. David Peat, *Turbulent Mirror* (New York: Harper, 1989), p. 140.

11. Briggs and Peat, *Turbulent Mirror*, p. 142.

12. John B. Cobb Jr., and Charles Birch, "Models of the Living," in *The Liberation of Life* (New York: Cambridge University Press, 1981), p. 91.

13. Briggs and Peat, *Turbulent Mirror*, p. 148.

14. Joseph Ford, "What Is Chaos, That We Should Be Mindful of It?" in *The New Physics*, Paul Davies, ed. (New York: Cambridge University Press, 1989), p. 348.

15. Whitehead, *Process and Reality*, p. 111.

Chapter 7

1. Sandra M. Faber, University of California, Santa Cruz, and Astronomer at Lick Observatory, private communication, 1999.

2. To better understand Planck time, we recall that massive particles have wavelike properties. It makes no sense to try to locate a particle in a distance smaller than its wavelength. According to the Heisenberg uncer-

tainty principle of quantum mechanics, the smallest wavelength a particle can have is when it is traveling close to the velocity of light. This is called its *Compton wavelength*.

If we consider a mass confined to a very small region, as the universe was in the beginning of the Big Bang, then it will form a black hole with a singularity and an event horizon, as discussed in chapter 2. If the event horizon is smaller than the Compton wavelength, then because of quantum mechanical uncertainty, we can't be certain whether the singularity is within or outside of the event horizon. In other words the general theory of relativity is no longer valid. Space and time are no longer smooth as the theory assumes—we have a sort of spacetime froth.

By equating the Compton wavelength and the event horizon of a given mass, we can find the mass of a black hole that can no longer be described by the general theory of relativity. This is the *Planck mass*. It is about 5×10^{-5} grams. The Compton wavelength of this mass is called the *Planck length*, about 4×10^{-33} centimeters. If we ask for the time for light to cross this distance, we find the Planck time, about 10^{-43} seconds.

3. Craig J. Hogan, "Primordial Deuterium and the Big Bang," *Scientific American* (December 1996): 68.

4. Malcolm Longair, "The New Astrophysics," in *The New Physics*, Paul Davies, ed. (New York: Cambridge University Press, 1989), p. 197.

5. Alan Guth and Paul Steinhardt, "The Inflationary Universe," in *The New Physics*, Paul Davies, ed. (New York: Cambridge University Press, 1989), p. 37.

6. Ibid., p. 35.

7. Frank Wilczek and Betsy Devine, *Longing for the Harmonies* (New York: Norton, 1988), p. 323.

8. Phillip J. E. Peebles, *Principles of Physical Cosmology* (Princeton, NJ: Princeton University Press, 1993), p. 423.

9. Wilczek and Devine, *Longing for the Harmonies*, p. 327.

10. Lawrence M. Krauss, "Cosmological Antigravity," *Scientific American* (January 1999): 53; Neta A. Bahcall, Jeremiah P. Ostriker, Saul Perlmutter, and Paul J. Steinbach, "The Cosmic Triangle: Revealing the State of the Universe," *Science* 284 (28 May 1999).

11. Craig J. Hogan, Robert P. Kirshner, and Nicholas B. Suntzeff, "Surveying Space-time with Supernovae, *Scientific American* (January 1999): 46.

12. Charles Hartshorne, *Creative Synthesis and Philosophic Method* (La Salle, ILL: Open Court Press, 1991), p. 210; Joe Rosen, "Response to Hartshorne Concerning Symmetry and Assymetry in Physics," *Process Studies* nos. 3–4 (1997): 318.

13. Penrose, *Emperor's New Mind*, p. 329.

14. E. Lerner, *The Big Bang Never Happened* (New York: Random, 1991); H. Arp, R. C. Keyes, and K. Rudnicki, *Progress in the New Cosmologies, Beyond the Big Bang* (Plenum Press, 1993). See also Geoffrey Burbridge, Fred Hoyle, and Jayant V. Narlikar, "A Different Approach to Cosmology," *Physics Today* (April 1999): 38, and rebuttal by Andreas Albrecht, "Reply to 'A Different Approach to Cosmology'," *Physics Today* (April 1999): 44.

15. Guth and Steinhardt, "Inflationary Universe," p. 45.

16. Ibid., p. 47.

17. Martin A. Bucher and David N. Spergel, "Inflation in a Low-Density Universe," *Scientific American* (January 1999): 63.

18. Stephen W. Hawking, *A Brief History of Time* (New York: Bantam Books, 1988), p. 117.

19. Ibid., p. 136.

20. Penrose, *Emperor's New Mind*, p. 351.

21. Paul Davies, *The Mind of God* (New York: Penguin Books, 1992), p. 68.

22. F. Flam, "COBE Finds Bumps in the Big Bang," *Science* (1 May 1992): 612.

23. Let us assume that a cubic centimeter of beach sand contains ten thousand grains. Then a cubic meter will contain a million times this or 10 billion grains (or 10^{10} grains). Assume a beach 1 kilometer long, 100 meters wide, and 1 meter deep. This beach contains 100,000 cubic meters of sand, and therefore 10^{15} grains of sand are on the beach. If we assume there are 100,000 kilometers of beaches in the world and each contains 100,000 cubic meters of sand, then the total number of grains of sand in all the beaches is 10^{20}. This is a factor of 100 less than the estimated 10^{22} stars in the universe.

24. G. B. Bothun, "The Ghostliest Galaxies," *Scientific American* (February 1997): 56.

25. S. E. Woosley and M. M. Philips, "Supernova 1987A!" *Science* 240 (6 May 1988): 750.

26. Penrose, *Emperor's New Mind*, p. 326.

27. Griffin, "Pantemporalism and Panexperientialism," in *The Textures of Time*, Paul A. Harris, ed. (Ann Arbor: University of Michigan Press, 1999), p. 36.

28. Ian G. Barbour, *Religion and Science, Historical and Contemporary Issues* (New York: Harper Collins, 1997), p. 212.

Chapter 8

1. Freeman Dyson, "Argument from Design," in *Disturbing the Universe* (New York: Harper, 1981), p. 250.

2. P. Sisterna and H. Vucetich, "Time Variation of the Fundamental Physical Constants: Bound from Geophysical and Astronomical Data," *Physical Review D* 41 (1991): 1034; Sisterna and Vucetich, "Time Variation of Fundamental Physical Constants. II Quark Masses as Time-Dependent Parameters," *Physical Review D* 44 (1992): 3096. See also A. Y. Potekhin and D. A. Varshalovich, "Non-variability of the Fine-Structure Constant Over Cosmological Time Scales," Astronomy and Astrophysics Supplement, ser. 104 (1994): 89–98.

3. Davies, *Mind of God*, p. 196.

4. J. Gribbin and Martin Rees, *Cosmic Coincidences* (New York: Bantam Books, 1989), p. 269.

5. Norman Pittenger, "Process Thought: A Contemporary Trend in Theology," in *Process Theology: Basic Writings by Key Thinkers of a Major Movement*, Ewert H. Cousins, ed. (New York and Toronto: Newman Press, 1971).

6. Whitehead, *Process and Reality*, p. 247.

7. Griffin, *A Process Philosophy of Religion*, (forthcoming).

8. Steven Weinberg, *The First Three Minutes* (New York: Basic, 1977), pp. 154–55.

9. Rebecca Parker, *Winter Solstice* (Berkeley, CA: Starr King School for the Ministry).

10. Whitehead, *Process and Reality*, p. 343.

11. Griffin, *A Process Philosophy of Religion*, (forthcoming).

12. Whitehead, *Adventures of Ideas*, p. 120.

13. *The Dialogues of Plato*, Benjamin Jowett, transl. (New York: Random, 1937), 2:255.

14. Whitehead, *Process and Reality*, p. 343.

15. Ibid., p. 344.

16. Griffin, *A Process Philosophy of Religion* (forthcoming) Chapter 5.

17. Whitehead, *Process and Reality*, p. 345.

18. Ibid., p. 346.

19. Ian Barbour, *Religion in an Age of Science* (San Francisco: Harper San Francisco, 1990), p. 160.

20. Marjorie H. Suchocki, *God, Christ, Church, A Practical Guide to Process Theology* (New York: Crossroad Publishing, 1982), p. 43.

21. Griffin, *Religion and Scientific Naturalism, Overcoming the Conflicts* (Albany: SUNY Press, 2000).

22. Palmyre M. F. Oomen, "The Prehensibility of God's Consequent Nature," *Process Studies* 27, nos. 1–2 (1988): 131.

23. Kim Chernin, *In My Father's Garden* (Chapel Hill, NC: Algonquin Books of Chapel Hill, 1996), p. 17.

24. Martin Buber, *Between Man and Man* (London: MacMillan, 1947), p. 10.

25. Whitehead, *Process and Reality*, p. 348.

26. Aldous Huxley, *The Perennial Philosophy* (New York: Harper, 1945), p. 59.

27. Cobb and Griffin, *Process Theology, An Introductory Exposition* (Philadelphia: Westminster Press, 1976), p. 8.

28. Edward Harrison, *Masks of the Universe* (New York: Macmillan, 1985), p. 266.

29. Whitehead, *Process and Reality*, p. 342.

30. Ibid.

31. R. Kane and S. H. Phillips, eds., *Hartshorne, Process Philosophy and Theology* (Albany: SUNY Press, 1989).

Chapter 9

1. Griffin, *A Process Philosophy of Religion*, (forthcoming).

2. *Process Studies*, published quarterly by the Center for Process Studies, Claremont, CA.

3. C. Robert Mesle, *Process Theology: A Basic Introduction* (St. Louis: Chalice Press, 1993), p. 62.

4. Charles Hartshorne, "The Rights of the Subhuman World," *Environmental Ethics* 1, no. 1 (spring 1979): 57

5. Mesle, *Process Theology*, p. 144.

6. John B. Cobb Jr. and Herman E. Daly, *For the Common Good* (Boston: Beacon, 1989), p. 209.

Glossary

Absolute temperature. The temperature that is measured from the point where molecular motion ceases. This point is about 273 degrees below zero on the centigrade scale. Absolute temperature is measured in degrees Kelvin, which is the temperature measured in degrees centigrade plus about 273.

Actual occasion. A unit of process. An actual occasion endures for but an instant, the instant of its becoming. It forms a unity from prehensions of the past and its goals and then becomes a datum for future actual occasions. According to Alfred North Whitehead they are "the final real things of which the world is made up."

Aggregate. A society of actual occasions without a dominant member, such as a rock.

Annihilation radiation. A gamma ray of 0.5 Mev energy (corresponding to the rest mass of an electron or positron) formed when an electron and positron change themselves into radiation upon contact.

Antimatter. An elementary particle that is an analogue of a given particle. It has the same mass and intrinsic angular momentum, but has the opposite electrical charge.

Atom. The smallest particle of an element that has the properties characterizing that element.

Baryon. An elementary particle formed of three quarks, for example, protons and neutrons.

Black hole. An object in which the gravity is so strong that even light cannot escape from it (the escape velocity exceeds the speed of light).

Boson. An elementary particle with a spin zero or an integral number of units of $\hbar$. Helium nuclei, photons, gravitons, and gluons are bosons.

The carriers exchanged to form the four known physical forces are bosons.

c. The symbol for the velocity of light in a vacuum, which is about three hundred thousand kilometers per second.

Color charge. The charge, analogous to an electric charge, which is present on quarks and gluons, and that is exchanged in the strong interaction. There are eight color charges.

Complex systems. Systems that involve many constituents in a web of interactions with new laws of behavior.

Compound individuals. A serially ordered society of actual occasions with a dominant member. The dominant occasion integrates the actual occasions into a higher-level conscious experience in the case of a human being.

Concrescence. The process in which an occasion gathers together its prehensions along with its subjective goals in order to produce a novel, present unity.

Consequent nature. The part of divinity that interacts with the world in process theology.

Cosmology. The study of the structure and evolution of the universe.

Dark matter. A cosmological puzzle resulting from the apparent lack of sufficient mass within the universe to account for its rate of expansion.

Dissipative systems. Systems that are far from equilibrium and that are sustained by an input of energy.

Electron. A pointlike particle that contains one unit of negative charge and that has a mass of 0.5 Mev/c^2. Electrons form the part of atoms outside the nucleus. As far as we know, the electron is not divisible.

Electroweak theory. A theory developed in the 1960s in which the electromagnetic theory of Maxwell is combined with weak interactions. This theory predicted the existence of W^t and Z^o bosons, which were subsequently discovered.

Enduring individual. A society that is ordered serially, such as a particle in a trajectory.

Entropy. A measure of the disorder in a system or alternately the energy in a system divided by its absolute temperature. According to the second law of thermodynamics, a closed system's entropy either stays constant or increases.

Event. A word that is synonymous with "occasion of experience" or "actual occasions," as used in this book.

Event horizon. The location around a black hole where the escape velocity is equal to the velocity of light; the surface of a black hole.

Exclusion principle. The requirement that two electrons (more generally two fermions) can't have the same quantum number.

Feeling. A positive prehension wherein a datum from a previous actual occasion is incorporated by a becoming occasion.

Fermion. An elementary particle that has spin 1/2, meaning it has an intrinsic angular momentum of $\hbar/2$. Electrons, nucleons, and quarks are fermions. Particles with spins of 3/2, 5/2 . . . are also fermions.

Field theory. The result of applying quantum mechanics to the behavior of a field, such as the electromagnetic field.

Flatness. The expanding universe may eventually collapse upon itself, expand forever with a finite velocity after an infinite time, or it may expand forever, but have zero velocity after an infinite time. The latter is termed a *flat universe*, since space will not be curved, either as a sphere or hyperboloid, but will be a flat plane. (If the cosmological constant is not zero, then the velocity of the universe will increase forever.)

Gamma (γ). The numerical factor in relativity that determines how much moving clocks slow down. It is numerically $1/\sqrt{1-v^2/c^2}$.

Galaxy. A large assemblage of stars, nebulas, and interstellar gas and dust.

Gluon. A virtual particle that is exchanged between quarks that constitutes the strong force. Gluons not only interact with quarks, but with each other.

Grand Unified Theory (GUT). A theory that describes and explains the four physical forces. This theory has not been found at this writing.

Graviton. A particle hypothesized that is the particle aspect of gravitational radiation.

Gravity waves. The wave aspect of gravitational radiation that is produced by accelerating masses.

$\hbar$. The basic quantum of angular momentum. It is Planck's constant divided by 2π and has the numerical value 7×10^{-22} *MeV*-seconds. The intrinsic angular momentum of an electron or nucleon in a particular direction is $\hbar/2$.

Heisenberg uncertainty relation. A fundamental postulate oı quantum mechanics that asserts that a pair of complementary variables such as position and momentum, or energy and time, cannot be measured with arbitrary precision.

Hubble sphere. The limit of our observable universe. This limit is reached when the expansion rate of the universe is equal to the velocity of light. Our Hubble sphere is about thirteen billion light years in radius.

Inflation. A hypothesis that just after quarks were formed in the Big Bang, there was an enormous expansion of the volume of the universe.

Initial aim. In process theology, it is the persuasive urge from the divine that has been tailored to enhance the intensity of experience for a particular occasion of experience or society of such occasions.

Isotope. The form of the same chemical element whose nuclei have the same number of protons but different numbers of neutrons.

Isotropy. The property of being the same in all directions.

Length contraction. An effect of the special theory of relativity in which a moving object appears contracted in its direction of motion with respect to an observer at rest.

Leptons. Spin 1/2 fermions without interactions involving the strong force. There are twelve known: the electron, muon, tauon, their associated neutrinos, and the antiparticles of each.

Mass-energy. The entity, according to the theory of relativity, of mass and energy that are each an aspect of a four-dimensional entity, mass-energy, so that they can be converted into each other.

Mental pole. The part of an occasion of experience that is internal and that synthesizes data from the physical pole with the occasion's subjective goals.

Meson. An elementary particle formed of two quarks.

Muon. A lepton similar to the electron but with a mass 207 times greater. The muon decays by the weak interaction with a half life of 2.2 microseconds.

Metaphysics. A framework of ideas that can be used to interpret all knowledge in a coherent, logical system.

Microwave background radiation. The radiation that is observed in every direction from outer space. It is presumably the cooled remnant radiation from the Big Bang.

Neutrino. The particles that have no electric charge and very little mass that are emitted in radioactive beta decay and in some other elementary particle processes. There are three kinds: electron neutrino, mu neutrino, and tau neutrino.

Neutron. An electrically neutral particle that is a constituent of nuclei of atoms along with protons. The neutron is formed from two down quarks and from one up quark.

Nucleon. A proton or a neutron, a constituent of the atomic nucleus.

Nucleus. The massive part of an atom, composed of protons and neutrons.

Objective immortality. Occurs when an event reaches satisfaction and becomes available as a datum for future events.

Occasion of experience. A spatiotemporal entity embodying a fundamental unit of experience from which more complex entities can be formed, sometimes called an actual occasion or event.

Openness. The availability of a choice among alternatives for an occasion of experience.

Panentheism. The theory that God is both immanent and transcendent. All is in God and God interacts with all.

Photon. The particle aspect of light, or more generally, electromagnetic radiation. Photons have no rest mass and carry one unit of angular momentum in units of. Each photon has an energy that is Planck's constant, h, multiplied by its frequency.

Physical pole. The part of the occasion of experience that receives input from the external world. It only receives data and makes no contribution of its own.

Positron. The antimatter particle of the electron. It has the same mass and intrinsic angular momentum, or spin, but has a positive charge rather than the electron's negative charge.

Precession. The rotation of the major axis of an elliptic orbit.

Prehends. A term in process philosophy that refers to the grasping or perceiving of a previous actual occasion by an occasion in the process of becoming. "Feels" is sometimes used to express this idea.

Prehension. The general way in which an occasion of experience in the process of becoming can include previous actual occasions and conceptual ideas.

Primordial nature. The part of the divine that is eternal and that is the source of possibilities for the creative advance in process theology.

Process theology. A theology of divinity that arises from process philosophy and that is necessary for the coherence of its metaphysics.

Proton. The nucleus of the hydrogen atom. It is formed from two up quarks and one down quark.

Pulsar. A star that produces regular periodic pulses of electromagnetic radiation.

QCD. Quantum chromodynamics. In analogy to QED, quantum electrodynamics, QCD, describes the interactions produced by color charges among quarks and gluons.

QED. Quantum electrodynamics, a very successful theory that combines electromagnetic theory and relativistic quantum mechanics. It describes the energy levels of the hydrogen atom that agree with experimentation to one part in ten billion and is used in many other calculations involving the microworld and electromagnetism.

Quark. A pointlike constituent of baryons, and mesons. The quarks contain all the nucleon mass and are indivisible as far as we know. Three quarks form a baryon, such as a proton, neutron, and other short-lived particles. Two quarks form mesons.

Satisfaction. The result of a concrescence in process philosophy. After its satisfaction an actual occasion becomes "objectively immortal" and is available to be prehended by future occasions.

Society. A complex of occasions of experience. The members of a society are alike, have a common element of form, and are self-sustaining. A particle in a trajectory forms a society of actual occasions, or a society of events; in this case it is a temporal society as one event succeeds another.

Spin. The intrinsic angular momentum possessed by elementary particles, such as electrons and protons.

Space-time. A new four-dimensional entity, which according to the theory of relativity, is when space and time become intermixed.

Strong force. The force produced by the exchange of gluons between quarks. It is responsible for holding nucleons together to form the nuclei of atoms.

Subjective goals. The plans for the future incorporated in the mental pole of an occasion of experience in process philosophy.

Substantialism. A worldview that unchanging substances are primary.

Supernova. An explosion of a star with a mass several times that of the sun. Supernovas form many of the more massive atoms in the periodic table and spew them into the universe.

Time dilation. An effect of the special theory of relativity in which moving clocks appear to run more slowly than clocks at rest with respect to the observer.

Virtual pairs. A pair of particles produced spontaneously in the vacuum as a result of the Heisenberg uncertainty principle, for example, an electron and positron can be produced and exist for a very short time arising out of nothing.

Wave function. The quantity that represents the state of a physical system in the Schrödinger formulation of quantum mechanics.

Weak force. The force involving the interactions of leptons. It is responsible for radioactive beta decay in which electrons are released spontaneously from nuclei.

***W*$^+$, *W*$^-$** bosons. Particles predicted by the electroweak theory that are exchanged to produce the weak force, and also have one unit of positive or negative electrical charge. To produce the weak force these particles are virtual, existing for a very short time.

***Z*^{o} boson.** A particle predicted by the electroweak theory that is exchanged in the weak interaction to form the weak force. It has no electric charge.

Suggested Additional Reading

Chapter 1. Process Philosophy

A. J. Bahm gives an introductory view of Buddhist philosophy in which we can see parallels to process thought. Ian Barbour places process theology in perspective with other religious views and with the world of science. David Ray Griffin's book A Process Philosophy of Religion gives a detailed and self-consistent treatment of process philosophy as it applies to religion. In Religion and Scientific Naturalism, Overcoming the Conflicts, Griffin demonstrates convincingly how process thought can bring together religion and science.

L. E. Hahn has edited over two dozen essays of Charles Hartshorne, who many consider the theological successor of Alfred North Whitehead. C. Robert Mesle's book is an excellent popular introduction to process thought. Modes of Thought provides additional insight into Whitehead's philosophy in addition to the references quoted in the "Notes." *The Philosophy of Nature* by Ivor Leclerc presents a fine historical view of the development of philosophies of nature leading to the development of process philosophy.

1. A. J. Bahm, *Philosophy of the Buddha* (Berkeley, CA: Asian Humanities Press, 1993)

2. Ian Barbour, *Issues in Science and Religion* (Englewood Cliff, NJ: Prentice Hall, 1966)

3. David Ray Griffin, *Religion and Scientific Naturalism, Overcoming the Conflicts* (Albany: SUNY Press, 2000).

4. David Ray Griffin, *A Process Philosophy of Religion* (forthcoming)

5. Lewis Edwin Hahn, ed., *The Philosophy of Charles Hartshorne* (LaSalle, ILL: Open Court, 1991)

6. Ivor Leclerc, *The Philosophy of Nature* (Washington: DC, Catholic University of America Press, 1989)

7. C. Robert Mesle, *Process Theology: A Basic Introduction* (St. Louis: Chalice Press, 1993)

8. Alfred North Whitehead, *Modes of Thought* (New York: Capricorn Books, 1958)

Chapter 2. Relativity

Lewis C. Epstein's Relativity Visualized is a nonmathematical book with many illustrations. Well-written undergraduate texts are Modern Physics by Raymond A. Serway, Clement J. Moses, and Curt A. Moyer and Concepts of Modern Physics, by Arthur Beiser. They also contain information concerning quantum mechanics and particle physics, which we shall discuss in subsequent chapters.

1. Arthur Beiser, *Concepts of Modern Physics* (New York: McGraw-Hill, 1987)

2. Lewis C. Epstein, *Relativity Visualized* (San Francisco: Insight Press, 1985)

3. Raymond A. Serway, Clement J. Moses, and Curt A. Moyer, *Modern Physics* (New York: Saunders College Publishing, 1989)

Chapter 3. The Microworld, Part I: Waves and Particles

For further reading on waves and particles in a popular vein the following books are recommended: Fritjof Capra's Tao of Physics and Gary Zukav's Dancing Wu Li Masters. Elementary undergraduate texts with excellent details are Modern Physics by Raymond A. Serway, Clement J. Moses, and Curt A. Moyer and Concepts of Modern Physics by Arthur Beiser.

1. Arthur Beiser, *Concepts of Modern Physics* (New York: McGraw-Hill, 1987)

2. Fritjof Capra, *The Tao of Physics* (Berkeley, CA: Shambala Press, 1975)

3. Raymond A. Serway, Clement J. Moses, and Curt A. Moyer, *Modern Physics* (New York: Saunders College Publishing, 1989)

4. Gary Zukav, *The Dancing Wu Li Masters* (New York: Morrow, 1979)

Chapter 4. The Microworld, Part II: Quantum Mechanics

A great number of books provide additional information on quantum mechanics. In a popular vein there are Roger Penrose's Emperor's New Mind and F. Wilczek and B. Devine's Longing for the Harmonies. Books with a mystical or metaphysical emphasis include K. Wilber's Quantum Questions and Nick Herbert's Quantum Reality. J. S. Polkinghorne's Quantum World provides a popular level but thorough account of the ideas of quantum mechanics. The author was a theoretical physicist, and is now an Anglican priest. Entry level college texts are Arthur Beiser's Concepts of Modern Physics and Modern Physics. Gary Zukav is a psychologist who is interested in physics. His book The Dancing Wu Li Masters provides the spirit of modern physics in a readable manner.

1. Arthur Beiser, *Concepts of Modern Physics* (New York: McGraw-Hill, 1987)
2. Nick Herbert, *Quantum Reality* (New York: Doubleday, 1985)
3. Roger Penrose, *The Emperor's New Mind* (New York: Oxford University Press, 1989)
4. J. C. Polkinghorne, *The Quantum World* (New York: Longman, 1984)

. Raymond A. Serway, Clement J. Moses, and Curt A. Moyer, *Modern Physics* (New York: Saunders College Publishing, 1989)

6. K. Wilber, ed., *Quantum Questions* (Boston: Shambala Press, 1985)
7. Gary Zukav, *The Dancing Wu Li Masters* (New York: Morrow, 1979)
8. Frank Wilczek and Betsy Devine, *Longing for the Harmonies* (New York: Norton, 1988)

Chapter 5. The Microworld, Part III: High-Energy Physics

For additional reading on elementary particle physics in a popular vein the reader is referred to Frank Wilczek and Betsy Devine's Longing for the Harmonies, G. Zukav's Dancing Wu Li Masters, and F. Capra's Tao of Physics. Current books at the undergraduate level are Modern Physics by Raymond A. Serway, Clement J. Moses, and Curt A. Moyer and Arthur Beiser's Concepts of Modern Physics. More advanced material that is quite current will be found in Particle Physics by B. R. Martin and G. Shaw and The New Physics by Paul Davies, ed.

1. Arthur Beiser, *Concepts of Modern Physics* (New York: McGraw-Hill, 1987)
2. Fritjof Capra, *The Tao of Physics* (Berkeley, CA: Shambala Press, 1975)

3. Paul Davies, ed., *The New Physics* (New York: Cambridge University Press, 1989)

4. B. R. Martin and G. Shaw, *Particle Physics* (New York: Wiley, 1982)

5. Raymond A. Serway, Clement J. Moses, and Curt A. Moyer, *Modern Physics* (New York: Saunders College Publishing, 1989)

6. Frank Wilczek and Betsy Devine, *Longing for the Harmonies* (New York: Norton, 1988)

Chapter 6. The Macroworld, Part I: Complex Systems

For further investigation into self-organizing and nonlinear systems the following books are recommended. Their themes concern time itself and involve the exploration of three centuries of science. Additional material concerning self-organization is available in an article by George Nicolis, a Belgian physical chemist.

In a popular vein: James Gleick's Chaos, with many fine illustrations, and Peter Coveney and Roger Highfield's Arrow of Time. Paul Davies's Cosmic Blueprint has a more philosophical approach. More technical, but fundamental books are those by Ilya Prigogine: From Being to Becoming, and Order Out of Chaos, and the article in The New Physics by Joseph Ford.

1. John Briggs and F. David Peat, *Turbulent Mirror* (New York: Harper, 1989)

2. Peter Coveney and Roger Highfield, *The Arrow of Time* (New York: Fawcett Columbine, 1990)

3. Paul Davies, *The Cosmic Blueprint* (New York: Simon & Schuster, 1988)

4. Joseph Ford, "What Is Chaos That We Should Be Mindful of It?" in *The New Physics*, Paul Davies, ed. (New York: Cambridge University Press, 1989)

5. James Gleick, *Chaos* (New York: Penguin Books, 1987)

6. George Nicolis, "Physics of Far-from-Equilibrium Systems and Self Organization," in *The New Physics*, Paul Davies, ed. (New York: Cambridge University Press, 1989)

7. Ilya Prigogine, *From Being to Becoming* (San Francisco: W. H. Freeman, 1980)

8. Ilya Prigogine, *Order Out of Chaos* (New York: Bantam Books, 1984)

Chapter 7. The Macroworld, Part 2: Cosmology

Adventures in Ideas contains a chapter that gives Alfred North Whitehead's perspective on nature and cosmology. Edward Harrison's book adopts an historical perpective with many physical insights as to the evolution of our concepts of the universe. Steven Weinberg's account in The First Three Minutes summarizes our knowledge from physics about what may have happened in the first few minutes after the Big Bang. The research briefing provided by the National Research Council gives us a thorough accounting of our cosmological knowledge as of 1995.

1. Edward Harrison, *Masks of the Universe* (New York: Macmillan, 1985)

2. David Schramm, *Cosmology, A Research Briefing*, Panel on Cosmology, National Research Council (Washington, DC: National Academy Press, 1995)

3. Steven Weinberg, *The First Three Minutes* (New York: Basic, 1988)

4. Alfred North Whitehead, *Adventures of Ideas* (New York: Macmillan, Free Press, 1933)

Chapter 8. Cosmology and Divinity

Ian Barbour places process theology in perspective with other religious views and the world of science in Religion and Science, Historical and Contemporary Issues. In A Process Philosophy of Religion David Ray Griffin gives a thorough treatment of the case for divinity from the process perspective—a naturalistic theism. In another book by Griffin Religion and Scientific Naturalism, Overcoming the Conflicts, the reader will find examples of harmonization of science and religion with a particularly detailed discussion of evolution. Lewis Edwin Hahn has edited over two dozen essays of Charles Hartshorne, who many consider the theological successor of Alfred North Whitehead.

C. Robert Mesle's book, Process Theology, has an excellent introduction to process theology as well as to a naturalist philosophy. The two books of Margaret Suchocki expand process theology from the Christian perspective. John B. Cobb Jr. and Griffin also give a fine introduction to process thought and later consider several aspects of it that relate to Christianity. Philip Clayton gives a detailed historical treatment of science and religion with a careful defense of panentheism, which is fundamental to process theology.

1. Ian Barbour, *Religion and Science, Historical and Contemporary Issues* (New York: Harper Collins, 1997)

2. David Ray Griffin, *A Process Philosophy of Religion* (forthcoming)

3. John B. Cobb Jr. and David Ray Griffin, *Process Theology, An Introductory Exposition* (Philadelphia: Westminster Press, 1976)

4. Lewis Edwin Hahn, ed., *The Philosophy of Charles Hartshorne* (LaSalle, ILL: Open Court, 1991)

5. C. Robert Mesle, *Process Theology, A Basic Introduction* (St. Louis: Chalice Press, 1993)

6. Margaret H. Suchocki, *The End of Evil* (Albany: SUNY Press, 1988)

7. Margaret H. Suchocki, *In God's Presence* (St. Louis: Chalice Press, 1996)

8. Philip Clayton, *God and Contemporary Science* (Edinburg, TX: Edinburg University Press, 1997)

Chapter 9. Epilogue: A World in Process

John B. Cobb Jr. has been a leader in process thought for several decades. With coauthor Herman E. Daly he develops a detailed critique of our economic system from the perspective of process thought in their book For the Common Good. In Sustainability he argues that it would be possible to provide for the needs of the present without endangering the resources of future generations if we adopt a process viewpoint. Modes of Thought give further views of Alfred North Whitehead on a variety of topics including his thoughts concerning nature and civilization.

1. John B. Cobb Jr., *Sustainability* (Maryknoll, NY: Orbis Books, 1995)

2. Herman E. Daly and John B. Cobb Jr., *For the Common Good* (Boston: Beacon, 1989)

3. Alfred North Whitehead, *Modes of Thought* (New York: Macmillan, 1938)

Note on Supporting Center

This series is published under the auspices of the Center for Process Studies, a research organization affiliated with the Claremont School of Theology and Claremont Graduate University. It was founded in 1973 by John B. Cobb, Jr., Founding Director, and David Ray Griffin, Executive Director; Marjorie Suchocki is now also a Co-Director. It encourages research and reflection on the process philosophy of Alfred North Whitehead, Charles Hartshorne, and related thinkers, and on the application and testing of this viewpoint in all areas of thought and practice. The center sponsors conferences, welcomes visiting scholars to use its library, and publishes a scholarly journal, *Process Studies*, and a newsletter, *Process Perspectives*. Located at 1325 North College, Claremont, CA 91711, it gratefully accepts (tax-deductible) contributions to support its work.

Index